Modelling of Floods in Urban Areas

Modelling of Floods in Urban Areas

Editors

Jorge Leandro
James Shucksmith

MDPI • Basel • Beijing • Wuhan • Barcelona • Belgrade • Manchester • Tokyo • Cluj • Tianjin

Editors

Jorge Leandro
Chair of Hydromechanics and
Hydraulic Engineering, Research
Institute of Water and
Environment
University of Siegen
Siegen
Germany

James Shucksmith
Department of Civil and
Structural Engineering
University of Sheffield
Sheffield
United Kingdom

Editorial Office
MDPI
St. Alban-Anlage 66
4052 Basel, Switzerland

This is a reprint of articles from the Special Issue published online in the open access journal *Water* (ISSN 2073-4441) (available at: www.mdpi.com/journal/water/special_issues/floods_urban).

For citation purposes, cite each article independently as indicated on the article page online and as indicated below:

LastName, A.A.; LastName, B.B.; LastName, C.C. Article Title. *Journal Name* **Year**, *Volume Number*, Page Range.

ISBN 978-3-0365-1620-2 (Hbk)
ISBN 978-3-0365-1619-6 (PDF)

© 2021 by the authors. Articles in this book are Open Access and distributed under the Creative Commons Attribution (CC BY) license, which allows users to download, copy and build upon published articles, as long as the author and publisher are properly credited, which ensures maximum dissemination and a wider impact of our publications.

The book as a whole is distributed by MDPI under the terms and conditions of the Creative Commons license CC BY-NC-ND.

Contents

About the Editors vii

Preface to "Modelling of Floods in Urban Areas" ix

Jorge Leandro and James Shucksmith
Editorial—Modelling of Floods in Urban Areas
Reprinted from: *Water* 2021, 13, 1689, doi:10.3390/w13121689 1

Binh Thai Pham, Mohammadtaghi Avand, Saeid Janizadeh, Tran Van Phong, Nadhir Al-Ansari, Lanh Si Ho, Sumit Das, Hiep Van Le, Ata Amini, Saeid Khosrobeigi Bozchaloei, Faeze Jafari and Indra Prakash
GIS Based Hybrid Computational Approaches for Flash Flood Susceptibility Assessment
Reprinted from: *Water* 2020, 12, 683, doi:10.3390/w12030683 5

Merhawi GebreEgziabher and Yonas Demissie
Modeling Urban Flood Inundation and Recession Impacted by Manholes
Reprinted from: *Water* 2020, 12, 1160, doi:10.3390/w12041160 35

Md Nazmul Azim Beg, Matteo Rubinato, Rita F. Carvalho and James D. Shucksmith
CFD Modelling of the Transport of Soluble Pollutants from Sewer Networks to Surface Flows during Urban Flood Events
Reprinted from: *Water* 2020, 12, 2514, doi:10.3390/w12092514 57

Esteban Sañudo, Luis Cea and Jerónimo Puertas
Modelling Pluvial Flooding in Urban Areas Coupling the Models Iber and SWMM
Reprinted from: *Water* 2020, 12, 2647, doi:10.3390/w12092647 75

Qing Lin, Jorge Leandro, Stefan Gerber and Markus Disse
Multistep Flood Inundation Forecasts with Resilient Backpropagation Neural Networks: Kulmbach Case Study
Reprinted from: *Water* 2020, 12, 3568, doi:10.3390/w12123568 91

Fenosoa Nanteneina Ramiaramanana and Jacques Teller
Urbanization and Floods in Sub-Saharan Africa: Spatiotemporal Study and Analysis of Vulnerability Factors—Case of Antananarivo Agglomeration (Madagascar)
Reprinted from: *Water* 2021, 13, 149, doi:10.3390/w13020149 111

Jaan H. Pu, Joseph T. Wallwork, Md. Amir Khan, Manish Pandey, Hanif Pourshahbaz, Alfrendo Satyanaga, Prashanth R. Hanmaiahgari and Tim Gough
Flood Suspended Sediment Transport: Combined Modelling from Dilute to Hyper-Concentrated Flow
Reprinted from: *Water* 2021, 13, 379, doi:10.3390/w13030379 135

Benjamin Dewals, Martin Bruwier, Michel Pirotton, Sebastien Erpicum and Pierre Archambeau
Porosity Models for Large-Scale Urban Flood Modelling: A Review
Reprinted from: *Water* 2021, 13, 960, doi:10.3390/w13070960 159

Detchphol Chitwatkulsiri, Hitoshi Miyamoto and Sutat Weesakul
Development of a Simulation Model for Real-Time Urban Floods Warning: A Case Study at Sukhumvit Area, Bangkok, Thailand
Reprinted from: *Water* 2021, 13, 1458, doi:10.3390/w13111458 173

About the Editors

Jorge Leandro

Prof. Dr. phil. habil Leandro leads the chair of hydraulics and hydromechanics, of the Forschungsinstitut Wasser und Umwelt at the University of Siegen. The chair has an own laboratory with a total of two test halls with an open area of 21 x 9 m (Hall I) and 12 x 8 m (Hall II), which are equipped with state-of-the-art measuring technology. Prof. Leandro holds the title of Agregado at the University of Coimbra, Portugal, and habil at the Technical University of Munich (TUM), Germany. These titles are only granted when significant outputs in terms of scientific activity, including also activities in postgraduation supervision and lecturing, have been achieved. He has 5 years' professional experience, and 10+ years of international academic experience. He is author of over 95 technical articles, with more than 50 publications in international journals (ISI). His research covers developments in the areas of Urban Flood Modeling and Forecasting, Flood Risk and Resilience Assessment, and Hydraulic.

James Shucksmith

Dr James Shucksmith is a Senior Lecturer of Water Engineering at the University of Sheffield. He was awarded his PhD on hydrodynamics and mixing processes in vegetated flows in 2008, and took up an academic position in 2010. His research interests include the physical modelling of urban flooding processes, water quality and transport processes in catchments and urban drainage networks, as well as the implementation of real time control systems. He has published over 30 papers in peer reviewed ISI indexed journal papers.

Preface to "Modelling of Floods in Urban Areas"

Understanding the risk of flooding in urban areas is a societal priority. However, there are significant technical challenges associated with the appropriate characterisation and representation of the numerous complex physical and hydrodynamic processes involved.

The aim of this Special Issue is thus to publish the latest advances and developments concerning the modelling of flooding in urban areas and contribute to our scientific understanding of the flooding processes and the appropriate evaluation of flood risk.

This issue contains contributions of novel methodologies including flood forecasting methods, data acquisition techniques, experimental research in urban drainage systems and sustainable drainage systems and new numerical approaches. It is addressed firstly to researchers, but practitioners will surely also find it very relevant.

In closing, we would like to acknowledge the work of the managing editor team, from the reviewers and over forty authors that submitted their work to this Special Issue.

We hope you enjoy reading it.

Jorge Leandro, James Shucksmith
Editors

Editorial

Editorial—Modelling of Floods in Urban Areas

Jorge Leandro [1,*] and James Shucksmith [2,*]

1. Chair of Hydromechanics and Hydraulic Engineering, Research Institute of Water and Environment, University of Siegen, Paul-Bonatz-Str. 9-11, 57068 Siegen, Germany
2. Department of Civil and Structural Engineering, University of Sheffield, Sheffield S1 3JD, UK
* Correspondence: jorge.leandro@uni-siegen.de (J.L.); j.shucksmith@sheffield.ac.uk (J.S.)

Citation: Leandro, J.; Shucksmith, J. Editorial—Modelling of Floods in Urban Areas. *Water* 2021, 13, 1689. https://doi.org/10.3390/w13121689

Received: 16 June 2021
Accepted: 17 June 2021
Published: 18 June 2021

Publisher's Note: MDPI stays neutral with regard to jurisdictional claims in published maps and institutional affiliations.

Copyright: © 2021 by the authors. Licensee MDPI, Basel, Switzerland. This article is an open access article distributed under the terms and conditions of the Creative Commons Attribution (CC BY) license (https://creativecommons.org/licenses/by/4.0/).

Understanding the risk of flooding in urban areas is a societal priority. The often-referenced challenges to flood prevention posed by climate change and rapid urbanization remain as pertinent as ever. Flood models have a variety of functions including real time warning, risk mapping, scenario evaluation, and the development of asset investment strategies. However, there are many significant technical challenges associated with the characterisation and representation of the numerous complex physical and hydrodynamic processes involved. Amongst others, these include rainfall-runoff, the heterogeneity of rainfall and surface topography at urban scales, and hydraulic interactions between overland and piped drainage systems. Recent advances in data driven techniques mean new approaches are increasingly becoming available alongside established hydrodynamic based methods, which can make real-time flood forecasting a possibility. Novel sensing, data acquisition systems, and experimental techniques also offer new opportunities for improved calibration, validation, and testing of flood models.

The aim of this special issue is thus to publish the latest advances and developments concerning the modelling of flooding in urban areas and contribute to our scientific understanding of the flooding processes and the appropriate evaluation of flood impacts. This issue contains contributions of novel methodologies including flood forecasting methods, data acquisition techniques, experimental research in urban drainage systems and/or sustainable drainage systems, and new numerical and simulation approaches in nine papers with contributions from over forty authors.

Selected highlights from each contribution are summarised as follows:

The paper "GIS Based Hybrid Computational Approaches for Flash Flood Susceptibility Assessment" [1] proposes and compares several novel hybrid computational approaches of machine learning methods for flash flood susceptibility mapping. About 320 past flash flood events and nine flash flood influencing factors, such as distance from rivers, aspect, elevation, slope, and land use were analyzed for the development of flash flood susceptibility maps. The results of this study suggested that the AdaBoostM1 based Credal Decision Tree has the best predictive capability in terms of accuracy.

The article "Modeling Urban Flood Inundation and Recession Impacted by Manholes" [2] introduces the flood inundation and recession model (FIRM) which is coupled to the commonly used SWMM urban drainage modelling package. FIRM computes the spread of surcharging water flow from a manhole based on the local topography evaluated using LIDAR elevation data. The model is validated using observations of a manhole overflow event in Edmunds, United States. Given the simplicity of the model, the paper highlights the potential further use of models of this type within real time or flood forecasting tools when considering flooding caused by surcharging manholes.

The editor choice article "CFD Modelling of the Transport of Soluble Pollutants from Sewer Networks to Surface Flows during Urban Flood Events" [3] utilises a 3D model of a surcharging urban drainage manhole structure in order to study the transport of soluble contamination from drainage networks into flood flows. The model is validated against an experimental dataset from a scaled physical model at the University of Sheffield, before being used to consider mixing processes and associated key timescales for pollutants to enter surface flows. The paper highlights further research opportunities in this area concerning the fate and transport of contaminants in urban flood water.

The featured article "Modelling Pluvial Flooding in Urban Areas Coupling the Models Iber and SWMM" [4] develops a free distribution dual drainage model linking the models Iber and Storm Water Management Model (SWMM). The dual drainage model links a 2D overland flow model and a 1D sewer network model with a Dynamic-link Library (DLL) that contains the functions in which the SWMM code is split. The developed model is particularly useful for urban areas, allowing the user to plan, evaluate and design new or existing urban drainage systems in a realistic way.

The editor choice article "Multistep Flood Inundation Forecasts with Resilient Back-propagation Neural Networks: Kulmbach Case Study" [5] presents an artificial neural networks (ANN) forecast framework for faster flood predictions. The framework is able to perform multi step forecasts for 1–5 h in a matter of seconds, triggered by a forecast threshold value. The ANN uses a high spatial resolution of 4 m $\times$ 4 m. For the historical flood events, the results show that the ANN outputs have a good forecast accuracy of the water depths for (at least) the 3 h forecast with over 70% accuracy, and a moderate accuracy for subsequent forecasts.

The featured article "Urbanization and Floods in Sub-Saharan Africa: Spatiotemporal Study and Analysis of Vulnerability Factors—Case of Antananarivo Agglomeration (Madagascar)" [6] performs a spatiotemporal analysis of the agglomeration of Antananarivo. It shows that urbanization leads to increased exposure of populations and constructions to floods. The study highlights that a share of the urban expansion in flood-prone zones is related to informal developments that gather highly vulnerable groups with very little in terms of economic resources. The authors suggest that an integration of flood risk management in spatial planning policies is an essential step to guide decisions in a sustainable way.

The feature article paper "Flood Suspended Sediment Transport: Combined Modelling from Dilute to Hyper-Concentrated Flow" [7] presents a modelling approach suitable for characterising the suspended sediment distribution within flood flows over a wide range of sediment concentrations. The model is parameterized and validated using a series of independent experimental laboratory datasets. The work highlights the opportunity to provide additional capability to flood flow modelling which may be relevant to health impact assessment and hazard evaluation.

The review paper "Porosity Models for Large-Scale Urban Flood Modelling: a Review" [8] considers recent developments in this specific approach to flood modelling. The review paper considers current and ongoing challenges associated with the effective parameterisation of different families of porosity models, and highlights ongoing work, for example, to improve the physical grounding of underlying modelling approaches and reduce mesh scale dependence of model parameters. A key recommendation is to establish suitable independent benchmark test cases for model testing, parameterisation, and evaluation.

The article "Development of a Simulation Model for Real-Time Urban Floods Warning: A Case Study at Sukhumvit Area, Bangkok, Thailand" [9] describes the development of a real time urban flood warning system deployed in the 24 sq.km case study area utilising 1D/2D dual drainage hydrodynamic modeling in conjunction with forecasted rainfall. The simulation is validated based on historic flooding records and, based on this analysis, the approach is shown to give a good representation of areas at risk. Based on an evaluation of previous rainfall events, the methodology can provide a flood warning lead time in the order of 10 min, which is limited by computational requirements.

This special issue highlights some of the ongoing challenges and the large variety of ongoing activity and techniques currently being used to model and understand flooding processes in urban catchments. We would highlight the variation of modelling techniques being used by the authors. In closing, we would like to acknowledge the work of the reviewers and all of the authors' submissions to this special issue. We hope you enjoy reading it.

Funding: This research received no external funding.

Institutional Review Board Statement: Not applicable.

Informed Consent Statement: Not applicable.

Data Availability Statement: No new data were created or analyzed in this study. Data sharing is not applicable to this article.

Conflicts of Interest: The authors declare no conflict of interest.

References

1. Pham, B.T.; Avand, M.; Janizadeh, S.; Phong, T.V.; Al-Ansari, N.; Ho, L.S.; Das, S.; Le, H.V.; Amini, A.; Bozchaloei, S.K.; et al. GIS Based Hybrid Computational Approaches for Flash Flood Susceptibility Assessment. *Water* **2020**, *12*, 683. [CrossRef]
2. GebreEgziabher, M.; Demissie, Y. Modeling Urban Flood Inundation and Recession Impacted by Manholes. *Water* **2020**, *12*, 1160. [CrossRef]
3. Beg, M.N.A.; Rubinato, M.; Carvalho, R.F.; Shucksmith, J.D. CFD Modelling of the Transport of Soluble Pollutants from Sewer Networks to Surface Flows during Urban Flood Events. *Water* **2020**, *12*, 2514. [CrossRef]
4. Sañudo, E.; Cea, L.; Puertas, J. Modelling Pluvial Flooding in Urban Areas Coupling the Models Iber and SWMM. *Water* **2020**, *12*, 2647. [CrossRef]
5. Lin, Q.; Leandro, J.; Gerber, S.; Disse, M. Multistep Flood Inundation Forecasts with Resilient Backpropagation Neural Networks: Kulmbach Case Study. *Water* **2020**, *12*, 3568. [CrossRef]
6. Ramiaramanana, F.N.; Teller, J. Urbanization and Floods in Sub-Saharan Africa: Spatiotemporal Study and Analysis of Vulnerability Factors—Case of Antananarivo Agglomeration (Madagascar). *Water* **2021**, *13*, 149. [CrossRef]
7. Pu, J.H.; Wallwork, J.T.; Khan, M.A.; Pandey, M.; Pourshahbaz, H.; Satyanaga, A.; Hanmaiahgari, P.R.; Gough, T. Flood Suspended Sediment Transport: Combined Modelling from Dilute to Hyper-Concentrated Flow. *Water* **2021**, *13*, 379. [CrossRef]
8. Dewals, B.; Bruwier, M.; Pirotton, M.; Erpicum, S.; Archambeau, P. Porosity Models for Large-Scale Urban Flood Modelling: A Review. *Water* **2021**, *13*, 960. [CrossRef]
9. Chitwatkulsiri, D.; Miyamoto, H.; Weesakul, S. Development of a Simulation Model for Real-Time Urban Floods Warning: A Case Study at Sukhumvit Area, Bangkok, Thailand. *Water* **2021**, *13*, 1458. [CrossRef]

Article

GIS Based Hybrid Computational Approaches for Flash Flood Susceptibility Assessment

Binh Thai Pham [1,*], Mohammadtaghi Avand [2,*], Saeid Janizadeh [2], Tran Van Phong [3], Nadhir Al-Ansari [4,*], Lanh Si Ho [5,*], Sumit Das [6], Hiep Van Le [1], Ata Amini [7], Saeid Khosrobeigi Bozchaloei [8], Faeze Jafari [2] and Indra Prakash [9]

1. University of Transport Technology, Hanoi 100000, Vietnam; hieplv@utt.edu.vn
2. Department of Watershed Management Engineering, College of Natural Resources, Tarbiat Modares University, Tehran 14115-111, Iran; Janizadeh.saeed@gmail.com (S.J.); Faeze_Jafari86@yahoo.com (F.J.)
3. Institute of Geological Sciences, Vietnam Academy of Sciences and Technology, 84 Chua Lang Street, Dong da, Hanoi 100000, Vietnam; tphong1617@gmail.com
4. Department of Civil, Environmental and Natural Resources Engineering, Lulea University of Technology, 971 87 Lulea, Sweden
5. Institute of Research and Development, Duy Tan University, Da Nang 550000, Vietnam
6. Department of Geography, Savitribai Phule Pune University, Pune 411007, India; sumit.das.earthscience@gmail.com
7. Kurdistan Agricultural and Natural Resources Research and Education Center, AREEO, Sanandaj 66177-15175, Iran; a.amini@areeo.ac.ir
8. Department of Watershed Management Engineering, College of Natural Resources, Tehran University, Tehran, 1417414418, Iran; saeid.khosro@yahoo.com
9. Department of Science & Technology, Bhaskarcharya Institute for Space Applications and Geo-Informatics (BISAG), Government of Gujarat, Gandhinagar 382007, India; indra52prakash@gmail.com
* Correspondence: binhpt@utt.edu.vn (B.T.P.); mt.avand70@gmail.com (M.A.); nadhir.alansari@ltu.se (N.A.-A.); hosilanh@duytan.edu.vn (L.S.H.)

Received: 11 January 2020; Accepted: 27 February 2020; Published: 2 March 2020

Abstract: Flash floods are one of the most devastating natural hazards; they occur within a catchment (region) where the response time of the drainage basin is short. Identification of probable flash flood locations and development of accurate flash flood susceptibility maps are important for proper flash flood management of a region. With this objective, we proposed and compared several novel hybrid computational approaches of machine learning methods for flash flood susceptibility mapping, namely AdaBoostM1 based Credal Decision Tree (ABM-CDT); Bagging based Credal Decision Tree (Bag-CDT); Dagging based Credal Decision Tree (Dag-CDT); MultiBoostAB based Credal Decision Tree (MBAB-CDT), and single Credal Decision Tree (CDT). These models were applied at a catchment of Markazi state in Iran. About 320 past flash flood events and nine flash flood influencing factors, namely distance from rivers, aspect, elevation, slope, rainfall, distance from faults, soil, land use, and lithology were considered and analyzed for the development of flash flood susceptibility maps. Correlation based feature selection method was used to validate and select the important factors for modeling of flash floods. Based on this feature selection analysis, only eight factors (distance from rivers, aspect, elevation, slope, rainfall, soil, land use, and lithology) were selected for the modeling, where distance to rivers is the most important factor for modeling of flash flood in this area. Performance of the models was validated and compared by using several robust metrics such as statistical measures and Area Under the Receiver Operating Characteristic (AUC) curve. The results of this study suggested that ABM-CDT (AUC = 0.957) has the best predictive capability in terms of accuracy, followed by Dag-CDT (AUC = 0.947), MBAB-CDT (AUC = 0.933), Bag-CDT (AUC = 0.932), and CDT (0.900), respectively. The proposed methods presented in this study would help in the development of accurate flash flood susceptible maps of watershed areas not only in Iran but also other parts of the world.

Keywords: machine learning; flash flood; GIS; Iran; decision trees; ensemble techniques

1. Introduction

Flash floods are those events where the rise in water is rapid within a few hours of the heavy rainfall. Flash flood is one of the most common, severely devastating natural hazards, which causes significant damages to the infrastructure and socioeconomy, and most importantly, it brings loss of lives [1–5]. Globally, more than 5000 people die each year due to flash flood events, which is about four times greater than any other category of flood event [6]. The most destructive nature of flood events is generally related to the extreme amount of torrential rainfall within a short duration resulting in high surface runoff [4,7]. Flash floods occur within catchments, where the response time of the drainage basin is short. According to the American Meteorological Society, flash flood events generally do not give advance warning and therefore, they cause significant risk and destruction due to their complex and dynamic environmental settings and nature [8,9].

Flash flood occurrence is affected by various watershed characteristics (type of basin and drainage), anthropogenic activities (land use, deforestation, and civil engineering construction) and meteorological conditions such as amount, intensity, spatial distribution, and time of rainfall. Recently, climate change is altering meteorological conditions which may lead to flash flood condition at one place and drought condition at another place. Therefore, the past may no longer be a reliable guide to the future. Thus, in the planning of flood management, especially of flash flood in urban areas, climate change effect is to be properly considered to avoid future damages to property and loss of life [10,11].

Geomorphological changes due to natural and anthropogenic causes can modify the flood pattern of different areas [12]. Urbanization is one of the important factors in the occurrence of flash floods in cities. Construction of roads and buildings reduces permeable areas and increases sealed surfaces (impermeable areas), thus causing less infiltration and more runoff with the same amount of rainfall causing pluvial flash floods [10]. Therefore, it is essential to identify and map accurately flash flood susceptible areas within a basin considering appropriate factors to develop suitable models for proper planning, management, and mitigation of flash flood events in an area [13].

There are many natural and anthropogenic factors that affect flood occurrence. Among these factors, topography is one of the important elements (land surface slope, river longitudinal profile, river cross section) that affects natural floods [14]. Flood parameters are very sensitive to topography changes. Low areas adjacent to rivers and streams have the highest risk of flooding. However, flash floods can also occur on hill slopes. Digital Elevation Model (DEM) as an indicator of the earth's surface contains information about the elevation of the earth. Flood depth and velocity are the most important parameters used in vulnerability assessment, estimation of casualties, and financial losses based on the land record [14]. Therefore, careful consideration of the topography of the area is desirable to avoid overestimation or underestimation of financial losses, casualties, and thus overall vulnerability assessment of an area [15,16].

Nowadays, multidisciplinary approaches including remote sensing, Geographic Information System (GIS), and machine learning methods are used for effective prediction and management of floods [5,6,12,17–19]. To recognize and delineate flash flood susceptible areas, DEM and other remote sensing satellite images have become popular and useful tools [20,21]. Bui and Hoang [22] reviewed the flash flood studies into three major classes, namely rainfall-runoff models, traditional methods, and pattern classification. In the case of rainfall-runoff models, the methodologies generally focus on establishing the relationship between the rainfall and runoff to determine the spatiotemporal distribution of the floods at a local scale and to carry out such studies in that area [23]. The traditional methods include analysis of long-term time series data and various statistical models [22]. The problem of predicting flash flood probability by implementing the above methods is the lack of reliable data availability of the long-term time series discharge records. Another method based on the pattern classification is relatively new, which employs monitoring of data at the gauging stations and also preparation of data of flooded and nonflooded group to assess the flash flood probability of a region and to demarcate the area where flash floods can occur [24,25].

Independent simplified decision-making techniques such as Analytical Hierarchy Process (AHP) [5,26–29], Fuzzy Logic (FL) [30,31], and Frequency Ratio (FR) [32,33] are some of the pattern classification-based methods which have been used to generate the flash flood maps around the world. Though these methods are simple, they do not provide a great level of accuracy in flash flood prediction in comparison to modern and advanced machine learning methods such as Support Vector Machine (SVM) [34,35], Artificial Neural Network (ANN) [36–38], Logistic Regression (LR) [39], GARP and QUEST [40], and Random Forest (RF) [41]. In recent years, some hybrid and ensemble machine learning methods such as Hybrid Bayesian Framework [24], Logistic Model Tree with Bagging Ensembles [42], Ensemble Weight-of-Evidence and Support Vector Machines [43], and Neuro-Fuzzy system integrated with Meta-Heuristic Algorithms [44] have been developed which provide better accuracy in comparison to single machine learning methods.

The main objective of the present study is to use GIS Based Hybrid Computational Approaches to develop ensemble models for accurate flash flood susceptibility assessment. In view of this, four hybrid ensemble models for the flash flood prediction were developed with Credal Decision Tree (CDT) as base classifier. These developed ensemble models are: AdaBoostM1 based CDT (ABM-CDT); Bagging based CDT (Bag-CDT); Dagging based CDT (Dag-CDT); and MultiBoostAB based CDT (MBAB-CDT). A small watershed of Tafresh county in the Markazi province of Iran, which experiences many flash floods every year, was selected as a study area for collecting and generating the datasets for the modeling process. To validate and compare performance of the models, various methods such as statistical measures and Area Under the Receiver Operating Characteristic (AUC) curve were used.

2. Materials and Methods

Description of the Research Area

Watershed of Tafresh county is one of the flash flood-prone areas of Markazi province. This county is located in the Markazi province of Iran covering an area of 1605 km^2, between 34°31′ N and 35°5′ N, 49°30′ E to 50°9′ E (Figure 1). Topography of the Tafrash watershed area is hilly with elevation ranging from 1296 to 3101 m. This area experiences cold winters and relatively moderate summers. The average temperature is 19.2 °C in summer and 6.4 °C in winter. Average annual rainfall in this region is 254.3 mm. Major water supply sources in the Tafresh watershed include springs, the perennial GharehChay River, the Ab Kamar seasonal river, and semi-deep wells. The GharehChay river with discharge 3000 ls^{-1} is one of the most important rivers in the area, which provides water for irrigation in Tafresh area, but due to droughts in recent years, discharge has reduced below 2000 ls^{-1}. However, several severe flash floods occur in the Tafrash watershed during winter every year, due to sudden heavy rainfall within a short period.

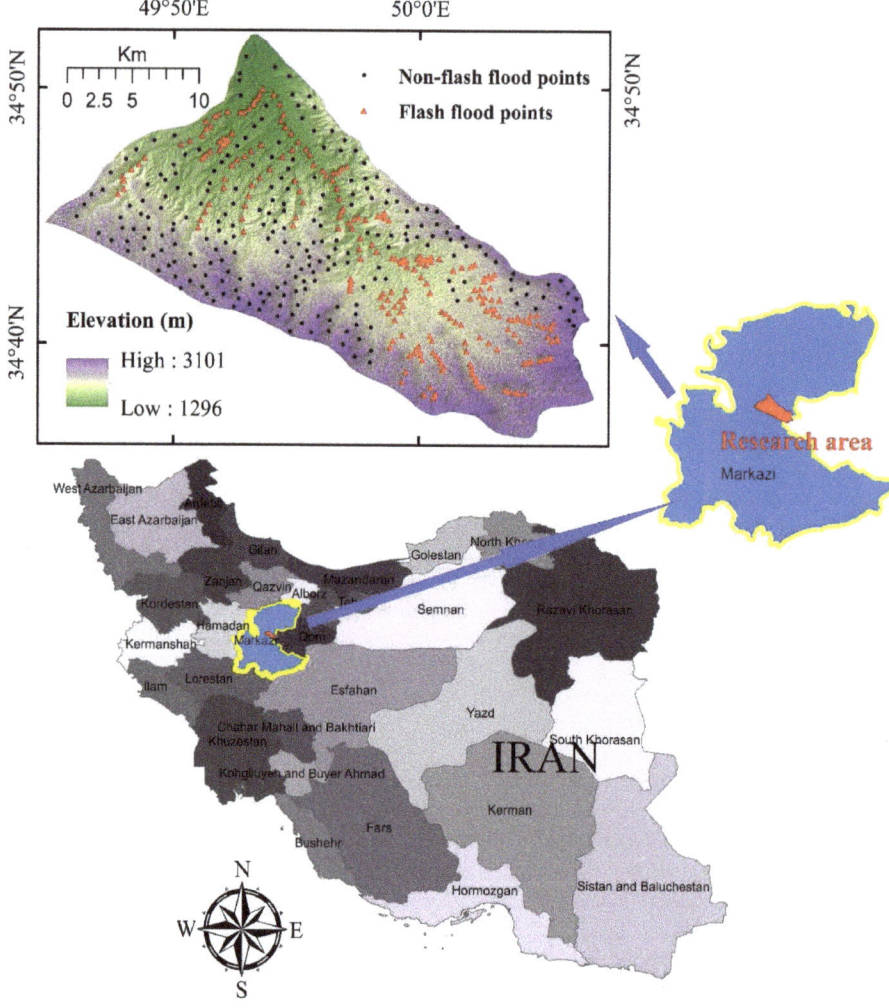

Figure 1. Location of study area.

3. Data Collection and Preparation

3.1. Flash Flood Inventory

Accurate mapping of the past flash flood events has a great impact on the accuracy of developed flash flood susceptibility maps. In order to predict the future flash flood events in a region, it is necessary to have records of the past flash flood events of the area [45]. These events depend on many factors including topography (terrain gradient), meteorology (antecedent rainfall), soil type, vegetative cover, and anthropogenic activities. These factors are considered as important parameters for the preparation of flash flood inventory and for the prediction of future flash flood events. In this research, in total, 320 past flash flood locations (represented on the maps by points) were obtained from the regional water organization of Markazi province (Figures 1 and 2). These flash flood points were divided randomly into 70% data points for training and 30% for validation purposes. In addition,

320 nonflooding points randomly selected from the high-altitude areas with low probability of flooding which were also used to combine with flash flood data for generating the training and testing datasets.

Figure 2. Flash flooding in Tafresh city.

3.2. Flash Flood Conditioning Factors

In this study, nine flash flood affecting parameters, namely distance from river, aspect, elevation, slope, rainfall, distance from faults, soil types, land use, and lithology were considered in the modeling. Thematic maps were generated using ArcGIS 10.1, ENVI 5.1, and SAGA-GIS 2 software (Figure 3). All these maps were converted to raster image (format) of 12.5 m × 12.5 m pixel size, which is up to the resolution of DEM for model studies (Table 1). A detailed description of these factors is given below:

Table 1. Data collection and preparation.

Row	Primary Input Data	Original Format Sources	Spatial Resolution	Source of Data	Derived Map
1	ALOS-PALSER DEM	Raster	12.5 m	https://search.asf.alaska.edu/	Slope, Aspect, Curvature, Elevation, Distance from river
2	Landsat 8 OLI	Raster	30 m	Department of Natural Resources of Markazi Province	Land use map
3	Meteorological data	Point	-	Markazi County Meteorological Bureau	Rainfall map
4	Geological map	Vector	1:100000	Geological survey and Mineral Exploration of Iran	Lithology and Distance from fault
5	Soil map	Vector	1:100000	Department of Natural Resources of Markazi Province	Soil map

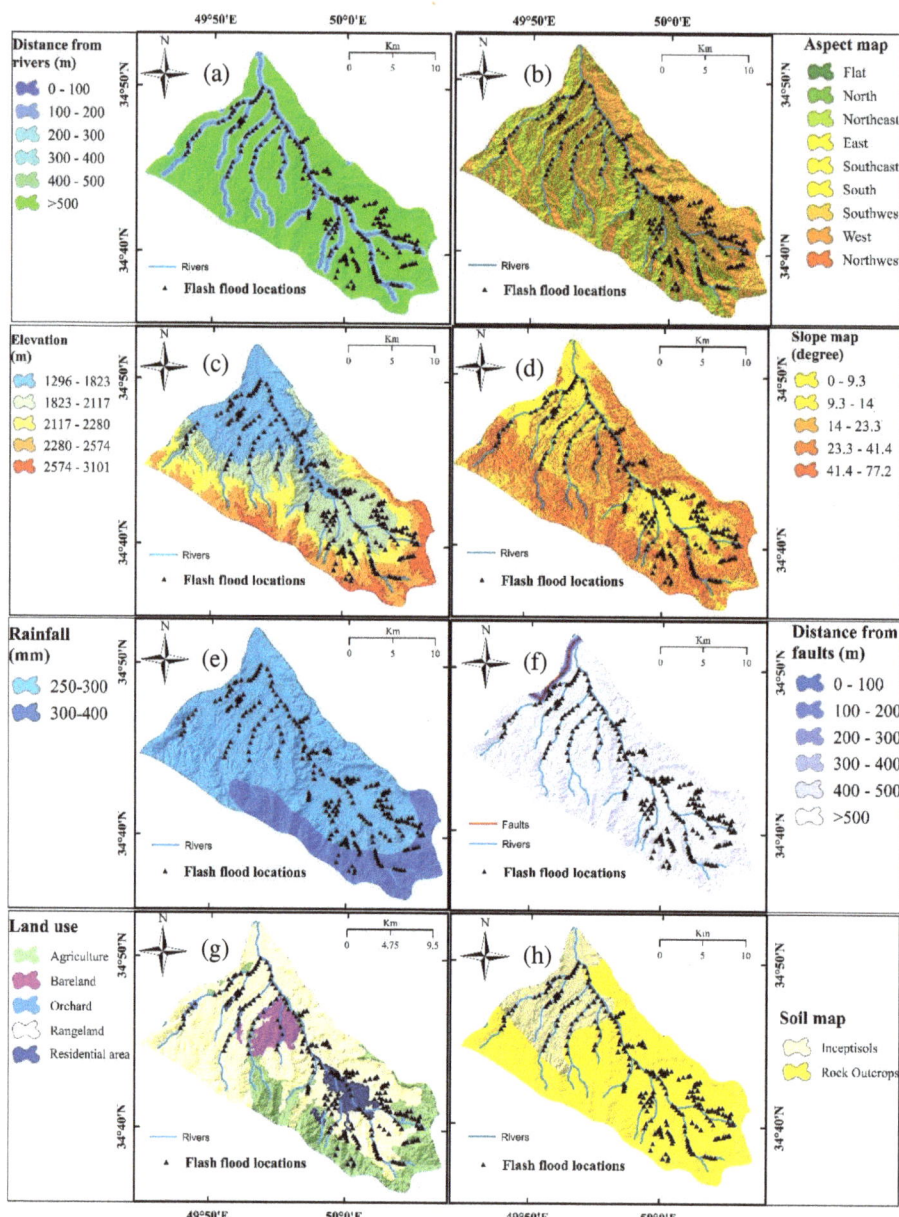

Figure 3. Cont.

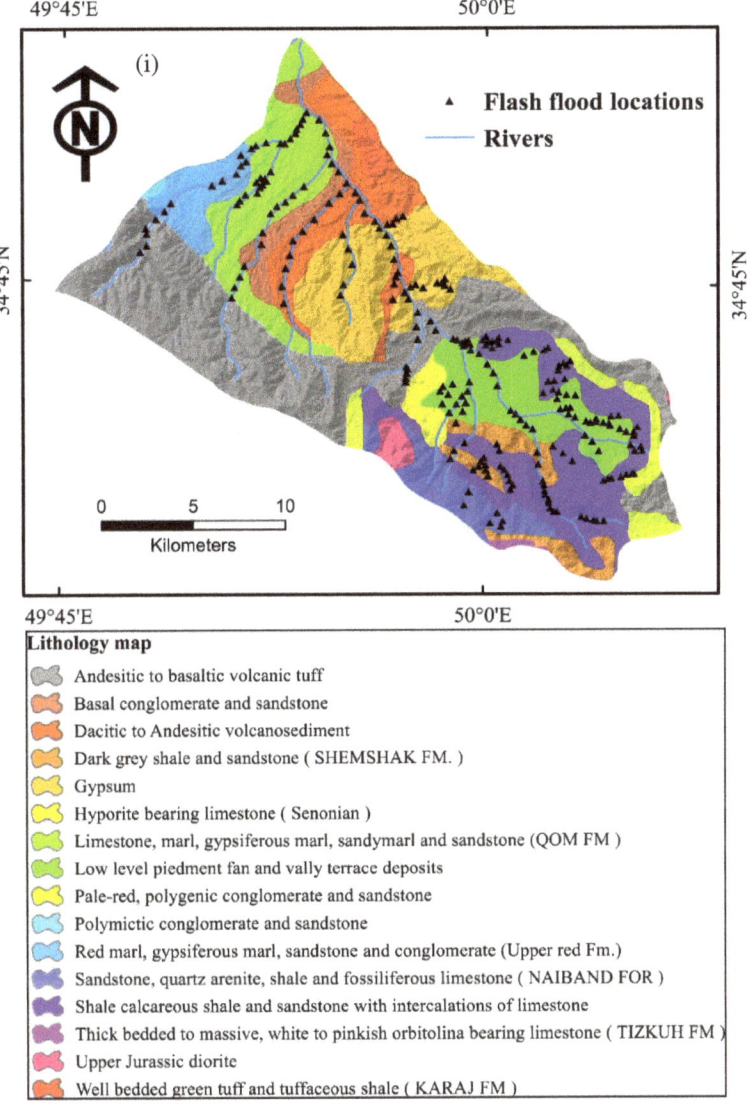

Figure 3. Maps of flash flood conditioning factors: (**a**) distance to rivers, (**b**) aspect, (**c**) elevation, (**d**) slope, (**e**) rainfall, (**f**) distance from faults, (**g**) land use, (**h**) soil, and (**i**) lithology.

Distance from rivers:

In general, the area which is close to the rivers is more prone to flooding in both cases of normal flood and flash flood within the river basin as water flows from higher elevation and accumulates at lower elevations. The areas close to other terrestrial water bodies such as ponds, dams, and lakes are also likely to be flooded in the event of heavy rains as the terrain in the vicinity of these water bodies would be almost flat [46]. However, pluvial flash floods may also occur at a distance away from the water bodies depending on the meteorological and topographical conditions. In the present study, six classes of buffer have been developed at buffer distance of 100 m from the river (Figure 3).

Aspect:

An aspect map of a region represents the direction of the surface slope. The direction which a slope faces with respect to the sun (aspect) has a profound influence on microclimate. The aspect map also shows no slope area (flat) where no surface slope is present; this is generally at the base of the hills or near lakes. Regions with low slope or regional flat surface are more vulnerable to the flash flood where water accumulates and rises [6,47]. Therefore, by using this parameter, the flat regions can easily be identified. Besides, flat area flooding also depends on the monsoon wind direction which hits the surface slope (Aspect). In this study, the aspect map was generated from the DEM with nine classes (Figure 3).

Elevation:

Water has a tendency of flowing from high altitude to lower elevation. The continuous flow of the rainwater therefore easily creates a flash flood situation in the low elevation areas [48,49]. However, pluvial flash floods also occur at higher elevation. In this study, an elevation map was generated from the DEM with five classes (Figure 3).

Slope:

Many factors affect catchment hydrologic characteristics, which ultimately influence the production of surface runoff. One of the important factors controlling runoff is surface slope [50]. On the steeper slopes, infiltration will be less and runoff will be more. This excessive runoff will cause flash flooding of the down slope flat areas. Thus, flat areas near and adjacent to high gradient slope generally have high probability of occurrence of flash floods [51]. In the present study, a slope map was created from DEM with five classes (Figure 3).

Rainfall:

Rainfall is the primary source of water for runoff generation over the land surface causing flooding of the low-lying areas. Runoff occurs whenever rain intensity exceeds the infiltration capacity of the ground (soil and jointed weathered rock). Intense short duration rainfall may cause flash floods. Rainfall is the most important factor for flooding of an area [50]. Flooding may also occur due to ice melting. In order to determine the annual rainfall map, the data of four rainfall-gauge stations for a period of 30 years were used. The rainfall map was divided into two classes (Figure 3).

Distance from faults:

Some of the major faults exposed on the surface with wide permeable fault zones may increase infiltration and thus reduce the runoff and can saturate surrounding groundmass causing local flooding. However, fault may also cause failure of levees and earthen dams due to structure failure and may result in flash flooding. In the present study, the distance from fault map was prepared into six classes (Figure 3).

Soil:

Soil is one of the important factors affecting infiltration and runoff and thus has a great impact on flooding. Soils rich in clay are mostly impermeable and cause more runoff and thus cause flooding of the area. In the present study, the soil map was developed from data obtained from the Soil Survey Department of Iran (Figure 3).

Land use:

Land use types affect the degree and frequency of floods in an area [52,53]. Infiltration and runoff depend on the land use pattern as well as other factors. Alterations in the land use configuration can change the flooding pattern of a region. Land-cover change due to anthropogenic activities such as urbanization, deforestation, and cultivation results in increased flash flood frequency and severity. In the present study, the land use map was obtained from the Department of Natural Resources of Markazi Province. Google Earth images and field survey were used to update the map (Figure 3).

Lithology:

Variation of lithology can strongly amplify or reduce the degree of flash flood vulnerability [54,55]. Infiltration and runoff depend on the permeability of lithounits as well as other geo-environmental factors. In this study, the lithology map was prepared from the Geological Survey of Iran data with sixteen groups (Table 2 and Figure 3).

Table 2. Lithology units in the Tafresh watershed and their relative permeability.

Group No	Geo-Units	Description	Permeability
1	Ea.bvt	Andesitic to basaltic volcanic tuff	Low
2	OMc	Basal conglomerate and sandstone	Moderate
3	Ed.avs	Dacitic to andesitic volcanosediment	Moderate
4	TRJs	Dark grey shale and sandstone (SHEMSHAK FM.)	Moderate
5	EKgy	Gypsum	High
6	K2l1	Hyporite bearing limestone (Senonian)	Moderate
7	OMq	Limestone, marl, gypsiferous marl. Sandymarl and sandstone (QOM FM)	Low
8	Qft2	Low level piedment fan and valley terrace deposit	High
9	Plc	Polymictic conglomerate and sandstone	Moderate
10	Mur	Red marl, gypsiferous marl, sandstone and conglomerate (upper red Fm.)	High
11	TRn	Sandstone, quartze arenite, shale and fossiliferous limestone (NAIBAND for)	Moderate
12	K2shm	Sale calcareous shale and sandstone with intercalations of limestone	Moderate
13	Ktzl	Thick bedded to massive, white to pinkish orbitolina bearing limestone (TIZKUh FM)	Moderate
14	Judi	Upper Jurassic diorite	Low
15	EK	Well bedded green tuff and tuffaceousshle (KARAJ FM)	Moderate

4. Methods Used

4.1. Frequency Ratio

Frequency Ratio (FR) determines the quantitative relationship between a flash flood event and its various variables [32,56]. In order to determine the FR, the ratio of flash flood events in each class of influencing factors is calculated relative to the total flash flood events. The ratio of the area of each class to the total area is also determined. Finally, by dividing the percentage of flash flood events in each class by the percentage of the area of each class relative to the entire research area, the FR of the classes of each factor is calculated. FR for each class of factors affecting the flash flood are calculated using the following equation [32,33]:

$$FR = \left(\frac{A}{B} \Big/ \frac{C}{D}\right) = \frac{E}{F} \quad (1)$$

where A: number of flash flood pixels per class, B: total flash flood pixels of the entire area, C: number of pixels per subclass of effective flash flood factors, D: total number of pixels in a region. E: percentage of flash flood occurrence in each class of effective factors, F: relative percentage of area of each class of total area.

4.2. Correlation Based Feature Selection

Irrelevant and redundant factors must be removed to improve data quality for modeling [57]. According to Pham et al. [58], working with a large number of factors reduces the speed of model execution, low modeling accuracy, and overfitting due to the large number of irrelevant factors as model inputs. There are many factors influencing the flood phenomenon, but the factors with higher correlation coefficients are more relevant in modeling and vice versa [58]. In this study, correlation based feature selection was selected to evaluate the importance of the factors used for better modeling of landslide susceptibility. This method is based on the assumption that features/factors are relevant if their values vary systematically with category membership [57,59]. In other words, a feature is useful if it is correlated with or predictive of the class; otherwise it is irrelevant [57,59]. In correlation based feature selection, the score of the evaluation is defined as Average Merit (AM) which is expressed as the following equation [57]:

$$AM_i = \frac{AC_i}{AI_i} \quad (2)$$

where AM_i is the score of factor ith, AC_i is the average correlation between the subset ith with the dependent variable, and AI_i is the average intercorrelation within the subset ith.

4.3. AdaBoostM1

AdaBoostM1 is a popular adaptive boosting algorithm proposed by Freund and Schapire [60]. AdaBoostM1 enhanced the predictive ability of the classifier. This method is employed to solve the classification problem, which contains a complicated dataset generated from previous classifiers. Firstly, the weight values are allocated to occurrences in learning dataset. After that, the weights are substituted in iterations of training process according to the performance of the previous base classifier. The training process will be terminated when the optimal weights have been given specifically to achieve the best performance of the base classifier [61].

4.4. Bagging

Bagging is known as one of the earliest ensemble methods which was proposed by Breiman [62] to improve the algorithm accuracy of machine learning methods [63]. In this method, bootstrap sample technique is used to produce numerous samples for creating a training classifier. Each generated training set is then employed to establish a decision tree. After that, these subsets are combined with the output in the final model [61]. This method not only enhances the capacity of generalization but also decreases the error of classification [64,65]. The optimum result of classification can be drawn using the following equation:

$$L'(x) = \underset{y \in Y}{\mathrm{argmax}} \sum_{i=1}^{t} 1(C_i(x) = y) \quad (3)$$

where $L'(x)$ expresses a combination of classifier and $C_i(x)$ denotes an indicator function.

4.5. Dagging

Dagging was initially introduced by Ting and Witten [66]. This method is recognized as one of the famous ensemble techniques. Aim of Dagging method is to improve accuracy in prediction of the classifier by combining varied samples of the training set [67]. A number of disjointed samples are employed rather than bootstrap samples to achieve the base classifier [66,68]. This method is a powerful technique for a single classifier, which has a poor time of complexity; thus, the outputs of algorithms with weak training are linked via the popular voting rule [67].

4.6. MultiBoostAB

MultiBoostAB is a combination ensemble learning algorithm, which is established on the basis of AdaBoostM1 and Wagging methods in order to hinder overfitting problem [69]. Wagging is a variable of Bagging, which exploits training cases using various weights that can reduce remarkably the bias of AdaBoostM1 technique [70]. Combinations of Wagging and AdaBoostM1 produce a framework that can transform a weak training classifier to a robust one. As MultiBoostAB is able to perform parallel processing, it is considered as a potential and computational method that has more advantages in comparison to Wagging and AdaBoostM1 methods [69]. MultiBoostAB method involves three main steps: (1) selection of a subset randomly from the original learning data and then to use it to produce fundamental classifier-based models; (2) the weights of occurrence are adjusted based on the predictive competence of the models; and (3) finally, new subsets from the occurrence weighting are chosen for training newer models [71].

4.7. Credal Decision Tree

Abellan and Moral originally proposed Credal Decision Tree (CDT) using an original split criterion that was built based on uncertainty measures as well as inaccurate probabilities [70]. CDT is used to tackle classification problems by employing credal sets [64,72,73]. In order to reduce generating a complicated decision tree in the building process of CDT, an exclusive criterion was introduced in case of the summation of uncertainties raising due to splitting, the construction process will stop [64,74].

In order to quantitatively evaluate the entire uncertainty of credal sets, an updated method was recommended based on the theory of Dempster and Shafer [75,76]. The function applied in measuring the total uncertainty is expressed in the following equation [77]:

$$EU(\chi) = NG(\chi) + RG(\chi) \qquad (4)$$

where EU expresses a value of entire uncertainty (i.e., total uncertainty), NG denotes a general nonspecificity function, and RG is a general randomness function for a credal set that represents a credal set. The successes and conclusions on the measurement of the summation of uncertainty were derived in previous literature of Abellan and Moral [78]. Besides, the detailed procedure for computing and properties of this measurement on EU were clearly described in previous studies [72,78]. To analyze probability intervals of individual variables, the inaccurate probability model was adopted [79,80]. Supposing that the Z is known as a variable that has values which are expressed by zj; then p(zj) is considered as the probability distribution, which reflects that each value of zj is determined as per the following formula [73,81]:

$$p(z_j) \in \left[\frac{m_{zj}}{M+r}, \frac{m_{zj}+r}{M+r} \right], \; j = 1, \ldots, k; \qquad (5)$$

where M and m_{zj} express the sample size and the event frequency (Z = z_j), respectively; and r is called the hyperparameter, which has values of 1 or 2, as stated by Walley [80].

4.8. Validation of the Models

Validation is important to determine the accuracy of the flash flood susceptibility models. To verify the prediction capability of models, it is desirable to assess as well as compare both learning and validating datasets [17,25,42]. In the present study, various validation criteria were adopted, namely Area Under the Receiver Operating Characteristic (ROC) curve and statistical measures.

4.8.1. Receiver Operating Characteristic (ROC) Curve

ROC curve is considered as a good tool for analyzing landslide and flood susceptibility models [17,42,82,83]. The x-axis of the ROC curve graph shows the specificity whereas the y-axis presents the sensitivity [84–88]. The area located under the ROC curve which is called the AUC is commonly employed to evaluate the prediction capacity of models [89–93]. Normally, the value of AUC has a range of 0.5–1.0 [94–96]. Higher value of AUC indicates better prediction capacity of the models [97–100]. The value of AUC is calculated by the following equation:

$$AUC = \frac{(\sum EC + \sum IC)}{(a+b)} \qquad (6)$$

where EC indicates the number of the accurately classified flash flood events, IC denotes the number of the inaccurately classified flash flood events, a is single flash flood event, and b is denotes the total number of flash flood events.

4.8.2. Statistical Measures

In the present study, seven popular statistical measures, namely Positive Predictive Value (PPV), Negative Predictive Value (NPV), Root Mean Square Error (RSME), Accuracy (ACC), Sensitivity (SST), Specificity (SPF), and Kappa index (k) were employed for assessing performance of the flash flood prediction models. The description of these indexes is summarized in Table 3.

Table 3. List of statistical measures employed in this research [101–105].

Statistical Measures	Formula
PPV (%)	$PPV = \frac{A}{A+B}$
NPV (%)	$NPV = \frac{C}{C+D}$
ACC (%)	$ACC = \frac{A+C}{A+C+B+D}$
SST (%)	$SST = \frac{A}{A+D}$
SPF (%)	$SPF = \frac{C}{C+B}$
k	$k = \frac{P_a - P_{est}}{1 - P_{est}}$ $P_a = (A + C)$ $P_{est} = (A + D) \times (A + D) + (B + C) \times (D + C)$

Where, A (true positive) and C (true negative) denotes the number of pixels of flash flood event classified correctly, whereas B (false positive) and D (false negative) are the numbers of pixels of nonflash flood event classified incorrectly. P_a and P_{est} are the measured and expected agreements, respectively.

RMSE is defined as the squared difference error between the model simulated and measured values. This method is popularly employed to assess flash flood susceptibility maps [17,42]. The smaller values of RMSE means the prediction capacity of the model is better. Determination of RMSE is calculated as follows [59,106–108]:

$$RMSE = \sqrt{\frac{1}{N} \cdot \sum_{i=1}^{L} (X_{model} - X_{act})^2} \qquad (7)$$

where X_{model} and X_{act} denote the model simulated and actual (i.e., measured) value, respectively; L stands for the summation of samples.

5. Methodology

Methodology of the study is presented below in several main steps: (1) Data collection and preparation; (2) Generating training and testing datasets; (3) Building the flash flood models; (4) Validation of the models; and (5) Generation of flash flood susceptibility maps (Figure 4). A more detailed description of these steps is given below:

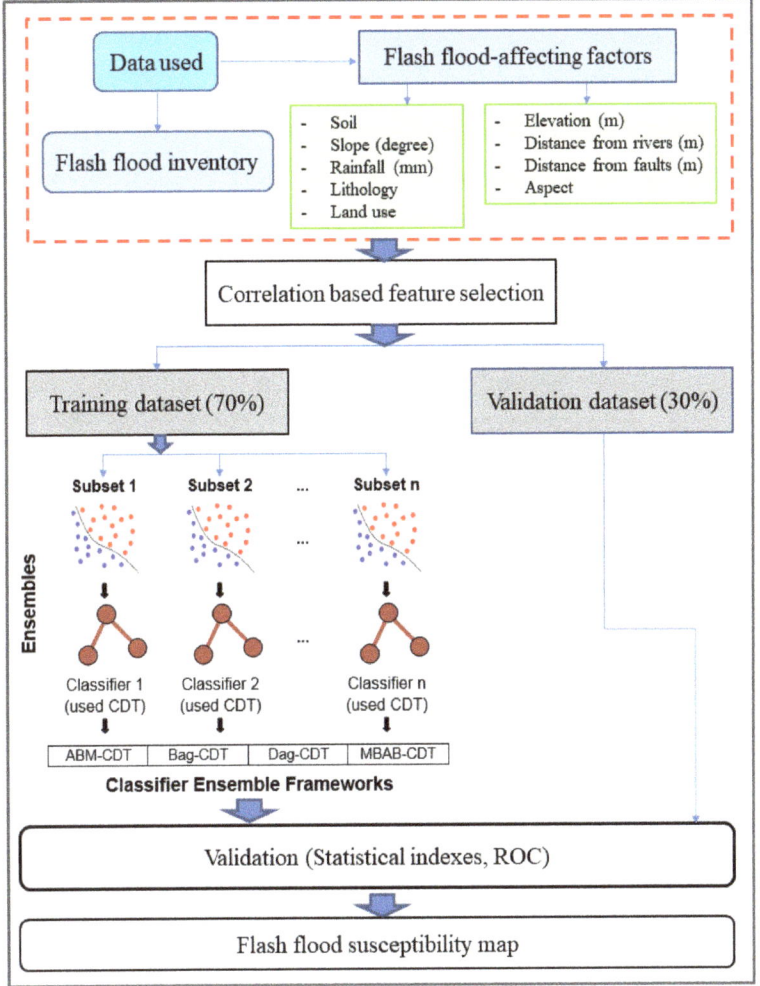

Figure 4. Methodological flowchart of the flash flood susceptibility mapping used in the study.

5.1. Data Collection and Preparation

Flash flood inventory map and the conditioning factor maps were generated in the raster format with 12.5 m pixel size. Thereafter, the inventory map was overlaid with the conditioning factor maps to calculate the FR values of each class of the conditioning factor using FR method. These FR values were then used as the weights of the class of the factors. In addition, correlation-based feature selection was used to validate and select the important factors and also to asses relative importance of these factors for modeling of flash floods.

5.2. Generating Training and Testing Datasets

Flash flood inventory was randomly divided into two parts with the ratio of 70/30. Out of these parts, 70% of inventory was used to sample with the conditioning factors assigned the weights for generating the training dataset, whereas the 30% remaining was used to sample with the conditioning factors assigned the weights for generating testing dataset. Selection of ratio for division of training and testing inventory might affect performance of the models. In this study, the ratio of 70/30 was used as it is a common ratio used in modeling [109–111]. This step was carried out in ArcGIS application.

5.3. Building the Flash Flood Models

Different hybrid models, namely ABM-CDT, Bag-CDT, Dag-CDT, MBAB-CDT, and a single classifier CDT were developed in this step using training dataset. Out of these methods, ABM-CDT is a combination of AdaBoostM1 ensemble and CDT classifier, Bag-CDT is a combination of Bagging ensemble and CDT, Dag-CDT is a combination of Dagging and CDT, and MBAB-CDT is a combination of MultiBoostAB and CDT. In these hybrid models, ensemble techniques were used to optimize the training dataset which was then used as input data in CDT classifier for flash flood susceptibility assessment. To construct these models, internal parameters should be selected and optimized to get the best performance of the models. More specifically, in CDT, initial parameters such as batch size, initial count, maximum of depth, minimum total weight of instances in a leaf, minimum proportion of variance, number of folds and seed were selected as 100, 0.0, −1, 2.0, 0.001, 3, 1, respectively. In ABM-CDT, initial parameters such as batch size, number of iterations, seed and weight of threshold were selected as 100, 10, 1, and 100, respectively. In Bag-CDT, initial parameters such as batch size, number of execution slots, number of iterations, and seed were selected as 100, 1, 15, and 1, respectively. In Dag-CDT, initial parameters such as batch size, number of folds, and seed were selected 100, 10, and 1, respectively. In MBAB-CDT, initial parameters such as batch size, number of iterations, number of subcommittees, seed, and weight of threshold were selected 100, 20, 3, 1, and 100, respectively. The values of these initial parameters of the models were determined by the trial-error process. This step was carried out using the packages and codes included in the Weka software.

5.4. Validation of the Models

Validation of the models was carried out on both training and testing datasets using various criteria such as PPV, NPV, SST, SPF, ACC, Kappa, RMSE, and AUC. While validation using training dataset shows the goodness-of-fit of the models, validation using testing datasets shows predictive capability of the models. This step was carried out using the packages and codes included in the Weka software.

5.5. Generation of Flash Flood Susceptibility Maps

In this step, flash flood susceptibility maps of Tafresh watershed were prepared based on ABM-CDT, Bag-CDT, Dag-CDT, MBAB-CDT hybrid machine learning models and CDT model in ArcGIS software. To construct the flash flood susceptibility maps, flash flood susceptibility indexes generated from the construction of the models were used to assign all pixels of the study area. Thereafter, these indexes were classified into five classes of flash flood susceptibility, namely very low, low, moderate, high, and very high to construct final maps using geometric interval classification method available in GIS software.

6. Results and Discussion

6.1. Impact Weight of each Class of Variables Affecting Flash Flood Susceptibility by FR Method

The impact weight of each class of variables was determined based on the comparative analyses of relationships between the location of past floods with the topographical and geo-environmental variables affecting flash flood occurrences (Figure 5). Analysis indicated that the highest weight in the variable of altitude classes belongs to the elevation class of 1296–1823 m. In the slope percentage of the surface slope, the weight of 0–9.3 degrees was the highest weight. In the slope direction variable, the northwest slope direction has a higher weight than the other aspects. In variable distance from the fault class of 400–500 m, weight has more influence than other classes. Examination of the variable distance from river showed that most of the flood-related weight was located at 0–100 m class. In the rainfall variable, the rainfall class 250–300 mm has higher weight than the other class. This means this class of rainfall belongs to threshold value for the occurrence of flash flood. Higher rainfall above this value can also cause flash flood depending on the duration in combination with other factors. Land use classes of the orchard and residential, which are in proximity to the main river and at gentle slopes, had the highest weighting factor compared to other land uses. Soil analysis indicates that the weight of the inceptisols soil is higher than that of the rocky outcrops. The lithology in this area indicates that Qom formation (OMq) has higher weight than other classes.

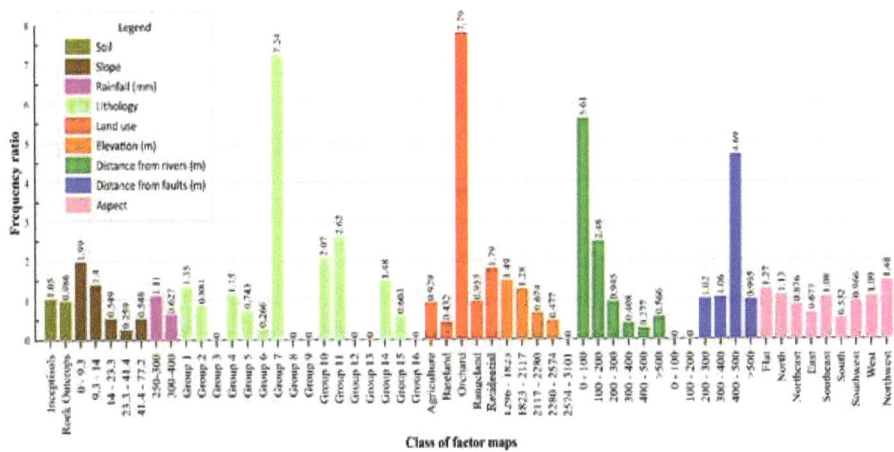

Figure 5. Frequency analysis of flash flood occurrence on the factor maps.

6.2. Importance of Factors Using Correlation-Based Feature Selection

Relative importance analysis of factors affecting flash floods was carried out using correlation-based feature selection method as shown in Table 4. It can be seen that distance from rivers is the most important factor for flash floods as the value of AM (0.608) is the highest compared with other factors. Following factors are slope (AM = 0.484), elevation (AM = 0.337), lithology (AM = 0.125), soil (AM = 0.099), rainfall (AM = 0.049), land use (AM = 0.024), aspect (AM = 0.022), and distance from faults (AM = 0.007), respectively. The feature selection results are reasonable as the areas close to the river are more likely to be affected by floods. This is true for normal river floods and also for flash floods in case of torrential rains within short period in this area [112,113]. Slope is also important as it influences surface runoff, volume, and velocity of flow. In the study area (Tafresh), there is more accumulation than outflow due to gentle topography (slope factor) resulting in the rapid rise of the flood water level within short time during torrential rain. Therefore, slope factor is the second most influential factor in the flood modeling (Table 4), which is consistent with many studies [114,115]. At higher elevation, slope factor is important resulting in higher velocity and runoff thus draining the water rapidly towards lower levels [116–118]. Other factors, namely lithology, soil, rainfall, land use, and aspect are also important factors for modeling of flash floods though their AM value varies as mentioned in Table 4. Here, we would like to mention that though AM of rainfall factor is only 0.049, it is the main and also triggering factor on which flash flood depends, especially in this area. However, the feature selection results show that distance to faults is the least important factor to flash flood occurrence and modeling (AM = 0.007), and thus this factor has a very small contribution to the performance of the models, and it should be removed from the datasets for further analysis of the models. Therefore, out of nine factors, only eight factors (distance from river, aspect, elevation, slope, rainfall, soil types, land use, and lithology) were reasonably selected for modeling of flash floods in this study.

Table 4. Importance of factors using correlation based feature selection.

Ranked	Class	Average Merit (*AM*)
1	Distance from rivers	0.608
2	Slope	0.484
3	Elevation	0.337
4	Lithology	0.125
5	Soil	0.099
6	Rainfall	0.049
7	Land use	0.024
8	Aspect	0.022
9	Distance from faults	0.007

6.3. Validation of Different Models

Performance of the machine learning models was validated using various criteria on both training and testing datasets (Figures 6–9). Validation of all the models was done by the ROC method (Figure 6). Results indicated very high AUC value during both training (ABM-CDT = 0.995; Bag-CDT = 0.972; Dag-CDT = 0.947; MBAB-CDT = 0.986; and CDT= 0.933) and testing phase (ABM-CDT = 0.96; Bag-CDT = 0.93; Dag-CDT = 0.47; MBAB-CDT = 0.933; and CDT = 0.90). Among all five models, ABM-CDT shows the maximum level of AUC compared with other models. All the models indicate a very low value of RMSE, both on the training dataset (ABM-CDT = 0.168; Bag-CDT = 0.245; Dag-CDT = 0.316; MBAB-CDT = 0.206; and CDT = 0.279) and testing dataset (ABM-CDT = 0.291; Bag-CDT = 0.307; Dag-CDT = 0.329; MBAB-CDT = 0.31; and CDT = 0.323) period, which clearly indicate high reliability of the proposed models (Figure 7). However, the ABM-CDT model indicates the best performance in comparison to other models, and it has the lowest RMSE value.

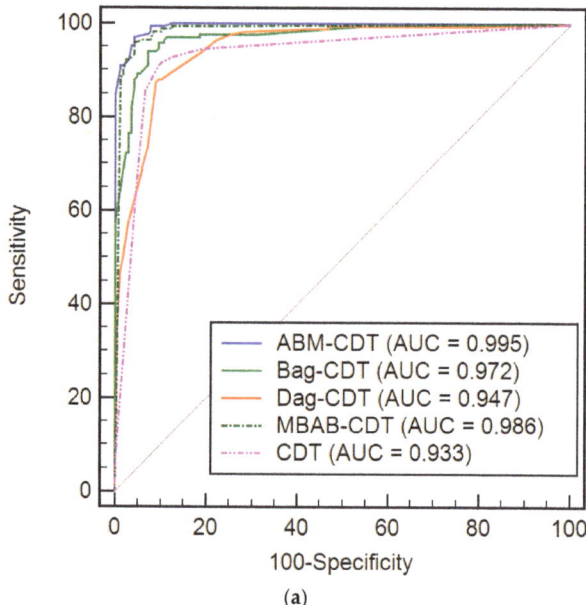

(a)

Figure 6. *Cont.*

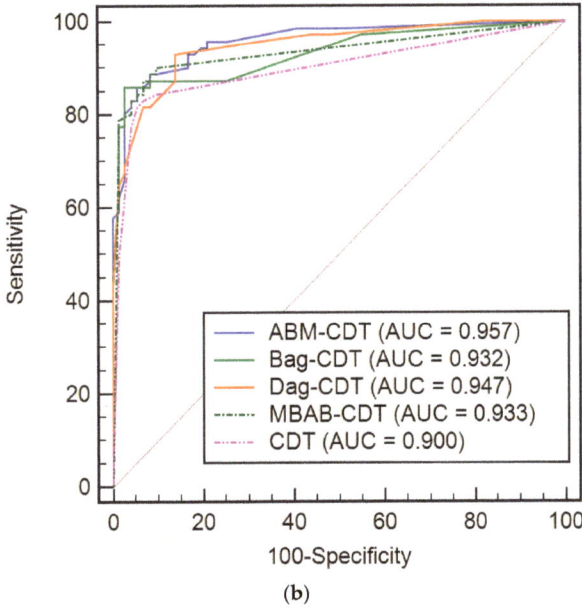

(b)

Figure 6. Analysis of Receiver Operating Characteristic (ROC) of the models: (**a**) training dataset and (**b**) validating dataset.

Figure 8 indicates performance of the models using other validation criteria. It can be observed that all models have good performance with high values of PPV, NPV, SST, SPF, and ACC. Out of these, the ABM-CDT model has high values of PPV (95.81% for training and 94.37% for testing), NPV (96.41% for training and 85.92% for testing), SST (96.39% for training and 87.01% for testing), SPF (95.83% for training and 93.85% for testing), and ACC (96.11% for training and 90.14% for testing), the Bag-CDT model has values of PPV (88.62% for training and 94.37% for testing), NPV (97.01% for training and 85.92% for testing), SST (96.73% for training and 87.01% for testing), SPF (89.5% for training and 93.85% for testing), and ACC (92.81% for training and 90.14% for testing), the Dag-CDT model has values of PPV (89.82% for training and 91.55% for testing), NPV (88.02% for training and 81.69% for testing), SST (88.24% for training and 83.33% for testing), SPF (89.63% for training and 90.63% for testing), and ACC (88.92% for training and 86.62% for testing), the MBAB-CDT model has values of PPV (92.22% for training and 94.37% for testing), NPV (96.41% for training and 84.51% for testing), SST (96.25% for training and 85.9% for testing), SPF (92.53% for training and 93.75% for testing), and ACC (94.31% for training and 89.44% for testing) and the CDT model has values of PPV (90.42% for training and 94.37% for testing), NPV (91.02% for training and 81.69% for testing), SST (90.96% for training and 83.75% for testing), SPF (90.48% for training and 93.55% for testing), and ACC (90.72% for training and 88.03% for testing). Kappa statistics also show a satisfactory accuracy in both the case of training (ABM-CDT = 0.922; Bag-CDT = 0.856; Dag-CDT = 0.788; MBAB-CDT = 0.898; and CDT = 0.814) and testing (ABM-CDT = 0.803; Bag-CDT = 0.803; Dag-CDT = 0.732; MBAB-CDT = 0.789; and CDT = 0.761) (Figure 9).

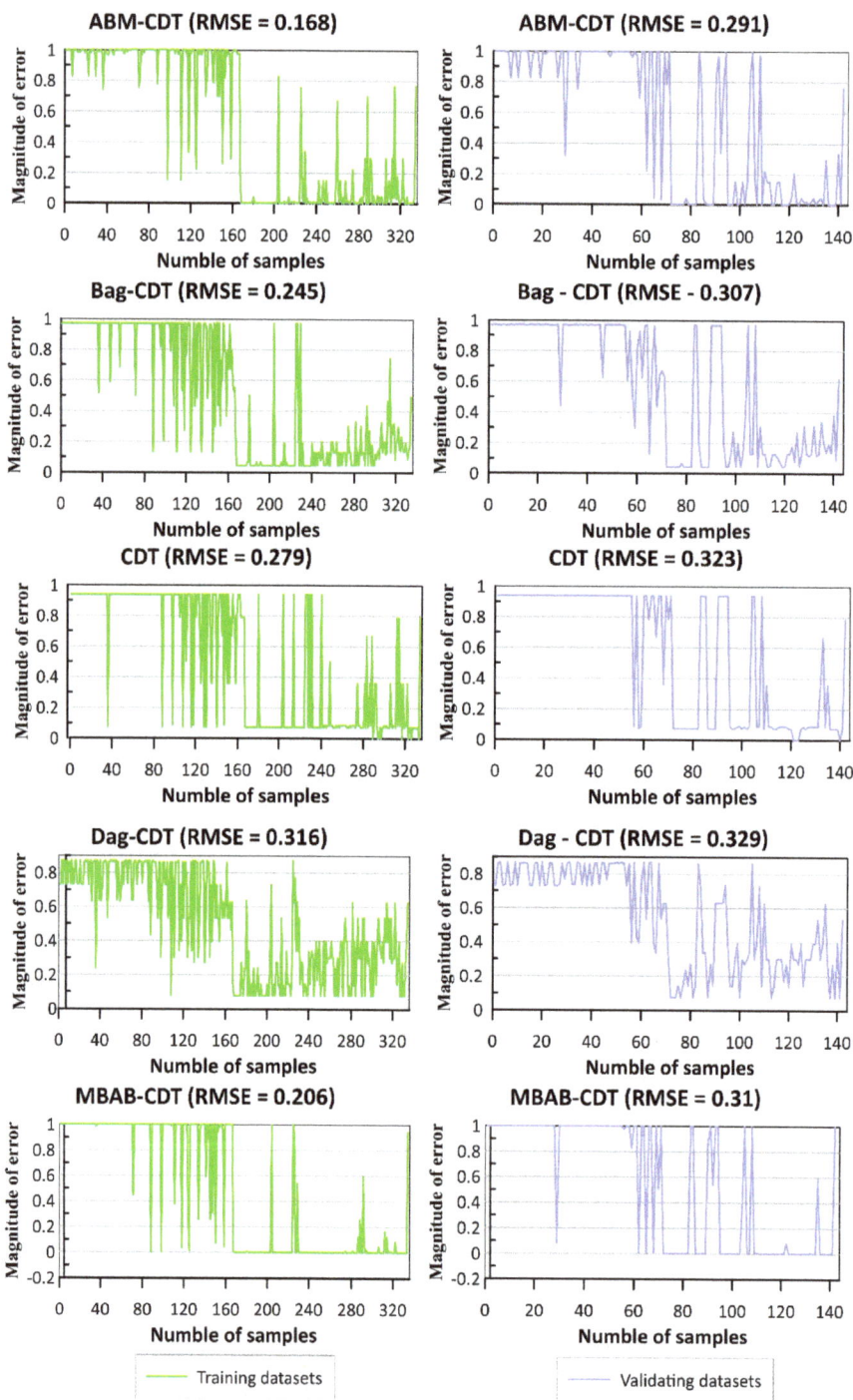

Figure 7. Analysis of RMSE of models.

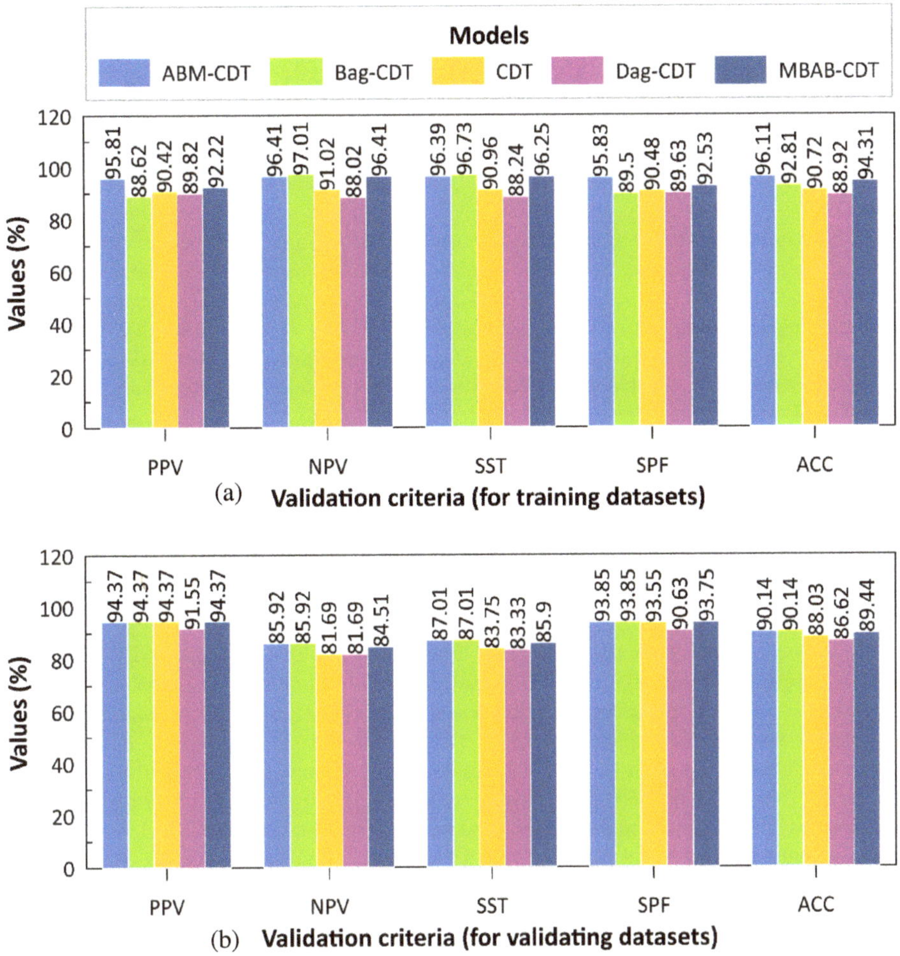

Figure 8. Analysis of accuracy of the models using: (**a**) training dataset and (**b**) validating dataset.

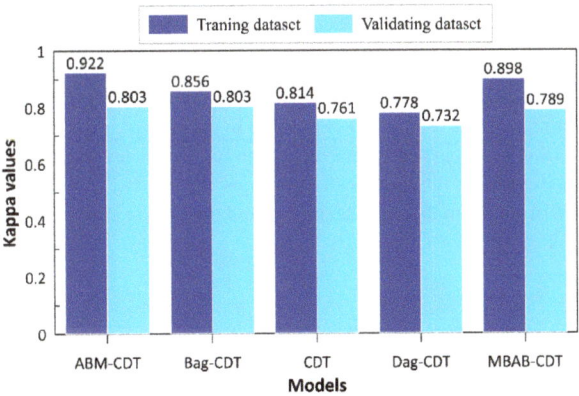

Figure 9. Kappa values for the models.

Considering analysis of the above results, it can be stated that all these developed and applied models performed well for flash flood susceptibility mapping in this study. In particular, the prediction capability of the CDT model has been enhanced by more than 5% with AdaBoost, about 3% with Bagging and MultiBoostAB, and 5% with Dagging. In general, CDT algorithm is one of the good data mining models built on the decision tree and uses IDM and general uncertainty measures [69]. However, it has a low accuracy as the built tree decides to categorize a new sample of data, especially with incomplete or missing values of the data. Therefore, the use of ensemble frameworks like AdaBoostM1, Bagging, Dagging, and MultiboostAB is a great help in improving performance of the CDT as these techniques have the capability to condense the bias as well as the variance and avoid the problem of overfitting [119]. Comparison results of different ensemble frameworks used in this study (ABM-CDT, Bag-CDT, Dag-CDT, and MBAB-CDT) showed that ABM-CDT outperforms other ensemble frameworks (Bag-CDT, Dag-CDT, and MBAB-CDT). Thus, it can be stated that AdaBoostM1 is more effective than other ensemble techniques (Dagging, Bagging, and MultiBoostAB) in improving performance of the CDT for flash flood susceptibility assessment of this study. This result is reasonable as AdaBoostM1 can be considered to make a classification of the binary classes and enhance the prediction accuracy [120,121]. It is a very well-known fact that among all these ensembles, AdaBoostM1 is an interpretable and highly robust algorithm that prevents noise in order to make significant improvement in classifying error in comparison to the base decision tree classifier [122]. Our results are comparable to the previous ensemble model-based studies, which report that the ensemble models lead to a boost in the performance of a standalone model [123–125].

6.4. Development of Flash Flood Susceptibility Maps

Flash flood susceptibility maps of the research area were produced using ABM-CDT, Bag-CDT, Dag-CDT, MBAB-CDT, and CDT models (Figure 10). Figure 11 shows the comparison of results of all the models of flash flood susceptibility classes and their percentage of class pixels and flash flood pixels. All the models indicated that more than 50% of past flash floods were observed on very high susceptibility class of the maps (ABM-CDT = 51.3%; Bag-CDT = 53.8%; CDT = 61.8%; Dag-CDT = 69.7%; and MBAB-CDT = 86.1%). Evaluation of the frequency ratio data of the historical flash flood locations and the generated flash flood maps for the very high susceptible pixel class was done. The maximum FR was observed for ABM-CDT (3.46) followed by Bag-CDT (3.44); Dag-CDT (3.4); CDT (2.88), and MBAB-CDT (2.66), which clearly indicated higher degree of reliability of ABM-CDT and Bag-CDT algorithms.

Analysis of the results of flash flood susceptibility maps shows that the Tafresh city area, which is located in the Tafresh watershed, belongs to very high susceptibility class. This is due to rapid development and expansion of the city area by encroaching topographically vulnerable areas to flash floods. Moreover, construction of buildings and roads in urban areas resulted in the increase of surface areas of impermeable structures and thus less infiltration and more runoff, causing flash floods in the event of intense rainfall during short periods [40,126,127]. The results of flood susceptibility zoning in Tafresh watershed showed that the southeastern parts have high to very high susceptibility to flash floods. The most important causes of flood susceptibility in these areas are related with anthropogenic activities causing drastic changes in catchment morphology, such as leveling of the land, altering the natural drainage, and increasing the impervious surfaces in the city. This has exacerbated the risk of floods and flooding of the infrastructure facilities thus increasing the potential threat to life and financial losses.

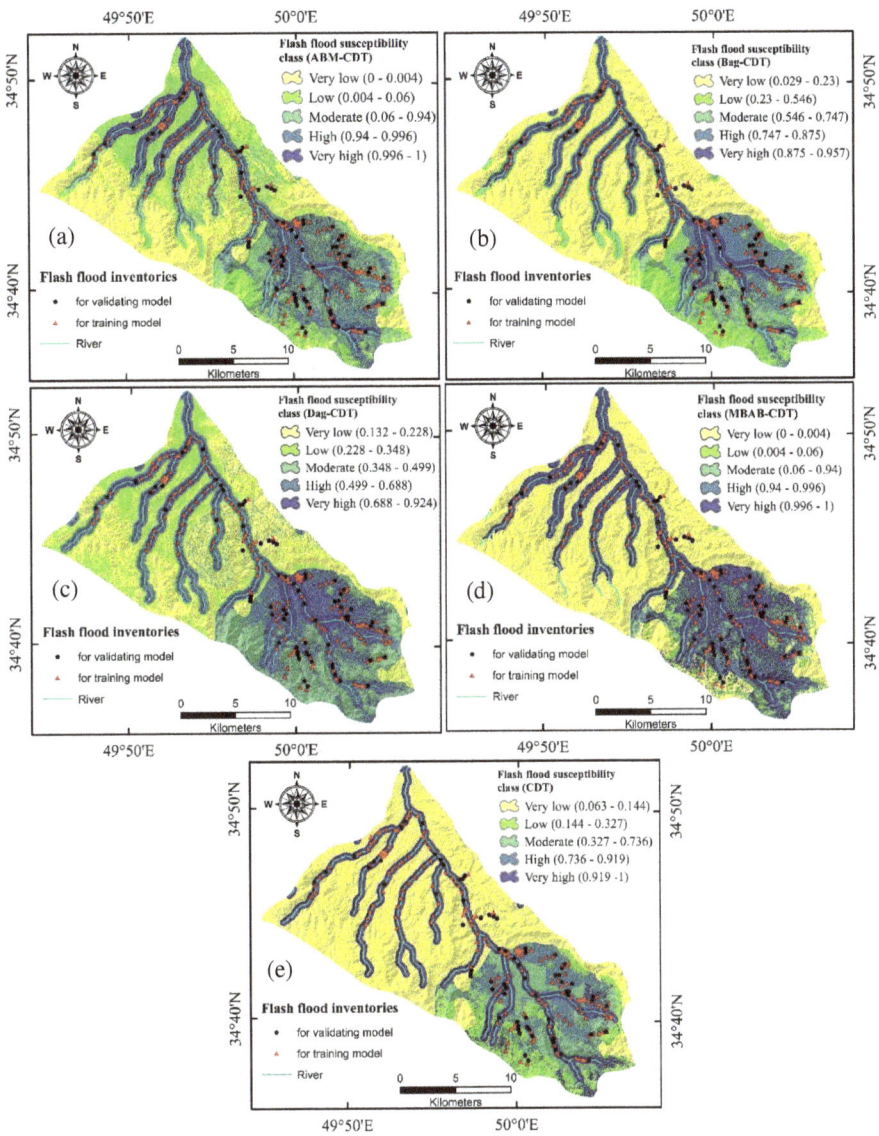

Figure 10. Flash flood susceptibility maps of the models: (**a**) ABM-CDT, (**b**) Bag-CDT, (**c**) Dag-CDT, (**d**) MBAB-CDT, and (**e**) CDT.

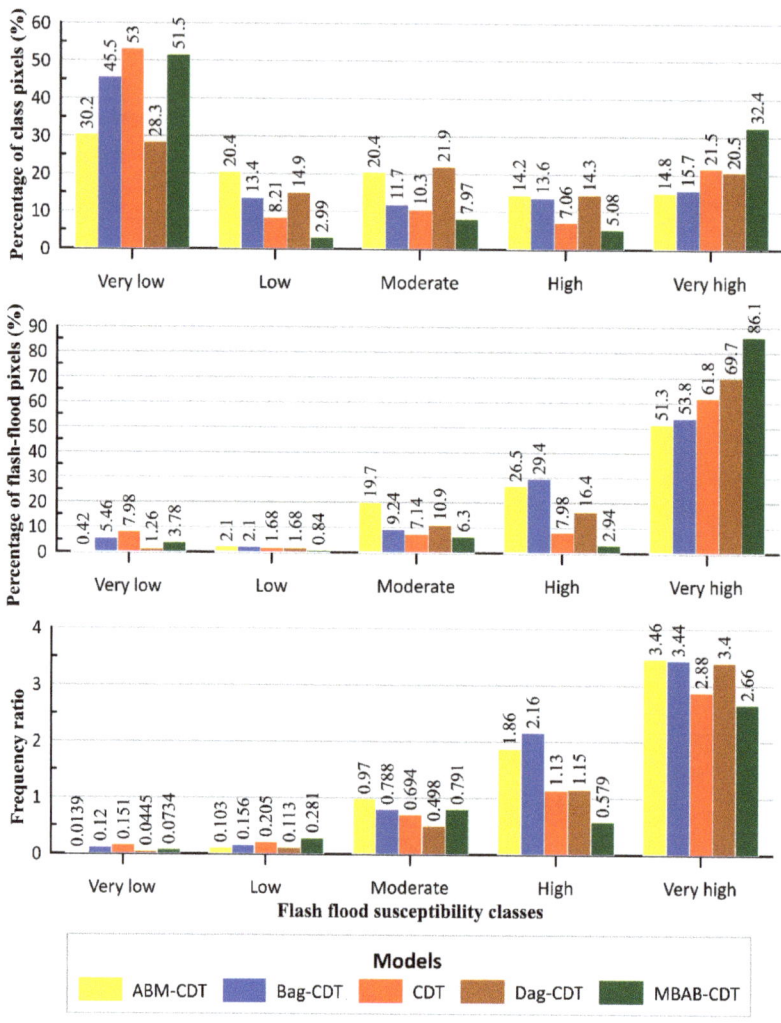

Figure 11. Analysis of performance of flash flood susceptibility maps using different models.

7. Concluding Remarks

In flash flood management studies, it is required to use accurate flash flood susceptibility maps by governing bodies and the policy makers for better flash flood mitigation and systematic development of the area. Since recent decades, a large number of methodologies have been developed to improve the accuracy of such maps. In this study, we proposed five new hybrid machine learning computational approaches to predict the possibility of flash flood occurrences in a studied catchment of Iran, where devastating flash flood events are frequent. The proposed methods are four hybrid models: ABM-CDT, Bag-CDT, Dag-CDT, MBAB-CDT, and single classifier: CDT. To construct the flash flood map, in total nine flash flood conditioning factors were taken into consideration to train and test the proposed models. Correlation based feature selection method was used to validate and select the important factors and also to asses relative importance of these factors for modeling of flash floods. Analysis shows that the lowest AM value (0.007) is of distance to fault and the highest AM value (0.608) is of distance to rivers. Distance to faults was then removed from the datasets for the flash flood modeling.

Therefore, in the present study, we have considered only eight factors (distance from river, aspect, elevation, slope, rainfall, soil types, land use, and lithology) in the modeling.

The results show that performance of all the studied models in terms of accuracy was good as these models show very low RMSE values and a high percentage of AUC. Results indicate very high AUC value during both training phase (ABM-CDT = 0.995; Bag-CDT = 0.972; Dag-CDT = 0.958; MBAB-CDT = 0.983; and CDT = 0.933) and testing phase (ABM-CDT = 0.96; Bag-CDT = 0.93; Dag-CDT = 0.95; MBAB-CDT = 0.933; and CDT = 0.90). Among all five models, ABM-CDT shows the maximum level of accuracy compared with other models. Evaluation of the FR data of the historical flash flood locations and generated flash flood maps was done for the very high susceptible pixel class. The maximum frequency ratio was observed for ABM-CDT (3.46), followed by Bag-CDT (3.44); Dag-CDT (3.5); CDT (2.88), and MBAB-CDT (2.65) which clearly indicated higher degree of reliability of ABM-CDT and Bag-CDT algorithms. The models, as an outcome of the study, would also help in the development of accurate flash flood susceptible maps in other watersheds of Iran. However, in the model studies, physical link between cause and effect is to be maintained considering local geo-environmental and hydrological factors for better flash flood prediction and management.

In this study, we performed a systematic analysis using multisource geospatial data; a significant number of limitations still exist in this study about data configuration. We have used 12.5 m spatial resolution ALOS-PALSER DEM which is freely available; a higher resolution DEM can provide a more reliable flood map which may be more useful for the practical use of flood mitigation. In addition, feature selection method such as Information Gain should be applied to evaluate the importance of input factors used for better investigation and application of the machine learning models. Furthermore, despite employing robust methodologies, our study area is local in nature. Therefore, this study is required to be extended to other places for the evaluation of its practical application in different terrains and environments.

In this study we did not consider dynamic changes which may be induced by human activities in the form of land use changes, topography alteration, infrastructure development, as well as climate change. These changes may affect the natural hydrological cycle and thus the pattern of floods, in particular of flash flood in urban areas impacting the life and property of communities affected. Another limitation of the model study is the lack of dynamic consideration of changing parameters related with physical changes, flow levels, direction, erosion, sedimentation, blocking of the drainage system, etc. on flood simulation and its causative effect on land development and flood management.

However, there is a great scope for further research related with the assessment, prediction, and mapping of flash floods by applying other combinations of hybrid artificial intelligence models in different areas using high resolution geo-spatial data for better production of flash flood susceptibility maps.

Author Contributions: Conceptualization, B.T.P., M.A., L.S.H., and N.A.-A.; data curation, S.J., F.J., and S.K.B.; methodology, N.A.-A., A.A., T.V.P., H.L.V., and B.T.P.; visualization, T.V.P., H.V.L., S.J., S.D., and S.K.B.; writing—original draft preparation, all authors; writing—review and editing, B.T.P., M.A., A.A., F.J., and N.A.-A.; supervision, N.A.-A., M.A., T.V.P., B.T.P. and I.P.; funding acquisition, B.T.P. and N.A.-A. All authors have read and agreed to the published version of the manuscript.

Funding: This research is funded by Vietnam National Foundation for Science and Technology Development (NAFOSTED) under grant number 105.08-2019.03.

Conflicts of Interest: The authors declare no conflict of interest.

References

1. Douben, K.-J. Characteristics of river floods and flooding: A global overview, 1985–2003. *Irrig. Drain. J. Int. Comm. Irrig. Drain.* **2006**, *55*, S9–S21. [CrossRef]
2. Anagnostou, M.N.; Kalogiros, J.; Anagnostou, E.N.; Tarolli, M.; Papadopoulos, A.; Borga, M. Performance evaluation of high-resolution rainfall estimation by X-band dual-polarization radar for flash flood applications in mountainous basins. *J. Hydrol.* **2010**, *394*, 4–16. [CrossRef]

3. Javelle, P.; Fouchier, C.; Arnaud, P.; Lavabre, J. Flash flood warning at ungauged locations using radar rainfall and antecedent soil moisture estimations. *J. Hydrol.* **2010**, *394*, 267–274. [CrossRef]
4. Modrick, T.M.; Georgakakos, K.P. The character and causes of flash flood occurrence changes in mountainous small basins of Southern California under projected climatic change. *J. Hydrol. Reg. Stud.* **2015**, *3*, 312–336. [CrossRef]
5. Das, S. Geospatial mapping of flood susceptibility and hydro-geomorphic response to the floods in Ulhas Basin, India. *Remote Sens. Appl. Soc. Environ.* **2019**, *14*, 60–74. [CrossRef]
6. Bui, D.T.; Tsangaratos, P.; Ngo, P.-T.T.; Pham, T.D.; Pham, B.T. Flash flood susceptibility modeling using an optimized fuzzy rule based feature selection technique and tree based ensemble methods. *Sci. Total Environ.* **2019**, *668*, 1038–1054. [CrossRef]
7. Georgakakos, K.P.; Hudlow, M.D. Quantitative precipitation forecast techniques for use in hydrologic forecasting. *Bull. Am. Meteorol. Soc.* **1984**, *65*, 1186–1200. [CrossRef]
8. Georgakakos, K.P. On the design of national, real-time warning systems with capability for site-specific, flash-flood forecasts. *Bull. Am. Meteorol. Soc.* **1986**, *67*, 1233–1239. [CrossRef]
9. Collier, C.G. Flash flood forecasting: What are the limits of predictability? *Q. J. R. Meteorol. Soc. J. Atmos. Sci. Appl. Meteorol. Phys. Oceanogr.* **2007**, *133*, 3–23. [CrossRef]
10. Recanatesi, F.; Petroselli, A.; Ripa, M.N.; Leone, A. Assessment of stormwater runoff management practices and BMPs under soil sealing: A study case in a peri-urban watershed of the metropolitan area of Rome (Italy). *J. Environ. Manag.* **2017**, *201*, 6–18. [CrossRef]
11. Szewrański, S.; Kazak, J.; Szkaradkiewicz, M.; Sasik, J. Flood risk factors in suburban area in the context of climate change adaptation policies—Case study of Wroclaw, Poland. *J. Ecol. Eng.* **2015**, *16*, 13–18. [CrossRef]
12. Tien Bui, D.; Khosravi, K.; Shahabi, H.; Daggupati, P.; Adamowski, J.F.; Melesse, A.M.; Pham, B.T.; Pourghasemi, H.R.; Mahmoudi, M.; Bahrami, S.; et al. Flood spatial modeling in northern Iran using remote sensing and gis: A comparison between evidential belief functions and its ensemble with a multivariate logistic regression model. *Remote Sens.* **2019**, *11*, 1589. [CrossRef]
13. Hammond, M.J.; Chen, A.S.; Djordjević, S.; Butler, D.; Mark, O. Urban flood impact assessment: A state-of-the-art review. *Urban Water J.* **2015**, *12*, 14–29. [CrossRef]
14. Saksena, S.; Merwade, V. Incorporating the effect of DEM resolution and accuracy for improved flood inundation mapping. *J. Hydrol.* **2015**, *530*, 180–194. [CrossRef]
15. Komolafe, A.A.; Herath, S.; Avtar, R. Sensitivity of flood damage estimation to spatial resolution. *J. Flood Risk Manag.* **2018**, *11*, 370–381. [CrossRef]
16. Annis, A.; Nardi, F.; Morrison, R.R.; Castelli, F. Investigating hydrogeomorphic floodplain mapping performance with varying DTM resolution and stream order. *Hydrol. Sci. J.* **2019**, *64*, 525–538. [CrossRef]
17. Khosravi, K.; Pham, B.T.; Chapi, K.; Shirzadi, A.; Shahabi, H.; Revhaug, I.; Prakash, I.; Bui, D.T. A comparative assessment of decision trees algorithms for flash flood susceptibility modeling at Haraz Watershed, Northern Iran. *Sci. Total Environ.* **2018**, *627*, 744–755. [CrossRef]
18. Nikoo, M.; Ramezani, F.; Hadzima-Nyarko, M.; Nyarko, E.K.; Nikoo, M. Flood-routing modeling with neural network optimized by social-based algorithm. *Nat. Hazards* **2016**, *82*, 1–24. [CrossRef]
19. Pradhan, B.; Shafiee, M.; Pirasteh, S. Maximum flash flood prone area mapping using RADARSAT images and GIS: Kelantan river basin. *Int. J. Geoinform.* **2009**, *5*, 11–23.
20. Noman, N.S.; Nelson, E.J.; Zundel, A.K. Review of automated floodplain delineation from digital terrain models. *J. Water Resour. Plan. Manag.* **2001**, *127*, 394–402. [CrossRef]
21. Papaioannou, G.; Vasiliades, L.; Loukas, A. Multi-criteria analysis framework for potential flash flood prone areas mapping. *Water Resour. Manag.* **2015**, *29*, 399–418. [CrossRef]
22. Bui, D.T.; Hoang, N.-D. A bayesian framework based on a gaussian mixture model and radial-basis-function fisher discriminant analysis (BayGmmKda V1. 1) for spatial prediction of floods. *Geosci. Model Dev.* **2017**, *10*, 3391.
23. Brunner, G.W. *HEC-RAS River Analysis System. Hydraulic Reference Manual, Version 1.0*; Hydrologic Engineering Center: Davis, CA, USA, 1995.
24. Bui, D.T.; Ngo, P.-T.T.; Pham, T.D.; Jaafari, A.; Minh, N.Q.; Hoa, P.V.; Samui, P. A novel hybrid approach based on a swarm intelligence optimized extreme learning machine for flash flood susceptibility mapping. *Catena* **2019**, *179*, 184–196. [CrossRef]

25. Bui, D.T.; Pradhan, B.; Nampak, H.; Bui, Q.-T.; Tran, Q.-A.; Nguyen, Q.-P. Hybrid artificial intelligence approach based on neural fuzzy inference model and metaheuristic optimization for flash flood susceptibilitgy modeling in a high-frequency tropical cyclone area using GIS. *J. Hydrol.* **2016**, *540*, 317–330.
26. Chen, Y.-R.; Yeh, C.-H.; Yu, B. Integrated application of the analytic hierarchy process and the geographic information system for flash flood risk assessment and flash flood plain management in Taiwan. *Nat. Hazards* **2011**, *59*, 1261–1276. [CrossRef]
27. Das, S. Geographic information system and AHP-based flood hazard zonation of Vaitarna Basin, Maharashtra, India. *Arab. J. Geosci.* **2018**, *11*, 576. [CrossRef]
28. Radwan, F.; Alazba, A.A.; Mossad, A. Flash flood risk assessment and mapping using AHP in arid and semiarid regions. *Acta Geophys.* **2019**, *67*, 215–229. [CrossRef]
29. Souissi, D.; Zouhri, L.; Hammami, S.; Msaddek, M.H.; Zghibi, A.; Dlala, M. GIS-based MCDM-AHP modeling for flash flood susceptibility mapping of arid areas, Southeastern Tunisia. *Geocarto Int.* **2019**, 1–27.
30. Pierdicca, N.; Pulvirenti, L.; Chini, M.; Guerriero, L.; Ferrazzoli, P. A fuzzy-logic-based approach for flash flood detection from cosmo-skymed data. In Proceedings of the IEEE International Geoscience & Remote Sensing Symposium, IGARSS 2010, Honolulu, HI, USA, 25–30 July 2010; pp. 4796–4798.
31. Zou, Q.; Zhou, J.; Zhou, C.; Song, L.; Guo, J. Comprehensive flash flood risk assessment based on set pair analysis-variable fuzzy sets model and fuzzy AHP. *Stoch. Environ. Res. Risk Assess.* **2013**, *27*, 525–546. [CrossRef]
32. Lee, M.-J.; Kang, J.; Jeon, S. Application of frequency ratio model and validation for predictive flooded area susceptibility mapping using GIS. In Proceedings of the 2012 IEEE International Geoscience and Remote Sensing Symposium, Munich, Germany, 22–27 July 2012; pp. 895–898.
33. Tehrany, M.S.; Pradhan, B.; Jebur, M.N. Flash flood susceptibility analysis and its verification using a novel ensemble support vector machine and frequency ratio method. *Stoch. Environ. Res. Risk Assess.* **2015**, *29*, 1149–1165. [CrossRef]
34. Yan, J.; Jin, J.; Chen, F.; Yu, G.; Yin, H.; Wang, W. Urban flash flood forecast using support vector machine and numerical simulation. *J. Hydro.* **2018**, *20*, 221–231. [CrossRef]
35. Tehrany, M.S.; Pradhan, B.; Mansor, S.; Ahmad, N. Flash flood susceptibility assessment using GIS-based support vector machine model with different kernel types. *Catena* **2015**, *125*, 91–101. [CrossRef]
36. Sahoo, G.B.; Ray, C.; De Carlo, E.H. Use of neural network to predict flash flood and attendant water qualities of a mountainous stream on Oahu, Hawaii. *J. Hydrol.* **2006**, *327*, 525–538. [CrossRef]
37. Youssef, A.M.; Pradhan, B.; Hassan, A.M. Flash flash flood risk estimation along the St. Katherine Road, Southern Sinai, Egypt using GIS based morphometry and satellite imagery. *Environ. Earth Sci.* **2011**, *62*, 611–623. [CrossRef]
38. Kia, M.B.; Pirasteh, S.; Pradhan, B.; Mahmud, A.R.; Sulaiman, W.N.A.; Moradi, A. An artificial neural network model for flash flood simulation using GIS: Johor River Basin, Malaysia. *Environ. Earth Sci.* **2012**, *67*, 251–264. [CrossRef]
39. Nandi, A.; Mandal, A.; Wilson, M.; Smith, D. Flash flood hazard mapping in Jamaica using principal component analysis and logistic regression. *Environ. Earth Sci.* **2016**, *75*, 465. [CrossRef]
40. Darabi, H.; Choubin, B.; Rahmati, O.; Torabi Haghighi, A.; Pradhan, B.; Kløve, B. Urban flash flood risk mapping using the GARP and QUEST models: A comparative study of machine learning techniques. *J. Hydrol.* **2019**, *569*, 142–154. [CrossRef]
41. Lee, S.; Kim, J.-C.; Jung, H.-S.; Lee, M.J.; Lee, S. Spatial prediction of flash flood susceptibility using random-forest and boosted-tree models in Seoul Metropolitan City, Korea. *Geomat. Nat. Hazards Risk* **2017**, *8*, 1185–1203. [CrossRef]
42. Chapi, K.; Singh, V.P.; Shirzadi, A.; Shahabi, H.; Bui, D.T.; Pham, B.T.; Khosravi, K. A novel hybrid artificial intelligence approach for flash flood susceptibility assessment. *Environ. Model. Softw.* **2017**, *95*, 229–245. [CrossRef]
43. Tehrany, M.S.; Pradhan, B.; Jebur, M.N. Flash flood susceptibility mapping using a novel ensemble weights-of-evidence and support vector machine models in GIS. *J. Hydrol.* **2014**, *512*, 332–343. [CrossRef]
44. Bui, D.T.; Panahi, M.; Shahabi, H.; Singh, V.P.; Shirzadi, A.; Chapi, K.; Khosravi, K.; Chen, W.; Panahi, S.; Li, S.; et al. Novel hybrid evolutionary algorithms for spatial prediction of floods. *Sci. Rep.* **2018**, *8*, 15364. [CrossRef] [PubMed]

45. Choubin, B.; Moradi, E.; Golshan, M.; Adamowski, J. An ensemble prediction of flood susceptibility using multivariate discriminant analysis, classification and regression trees, and support vector machines. *Sci. Total Environ.* **2019**, *651*, 2087–2096. [CrossRef] [PubMed]
46. Reager, J.T.; Thomas, B.F.; Famiglietti, J.S. River basin flash flood potential inferred using grace gravity observations at several months lead time. *Nat. Geosci.* **2014**, *7*, 588. [CrossRef]
47. Hoang, L.P.; Biesbroek, R.; Tri, V.P.D.; Kummu, M.; Van Vliet, M.T.H.; Leemans, R.; Kabat, P.; Ludwig, F. Managing flash flood risks in the mekong delta: How to address emerging challenges under climate change and socioeconomic developments. *Ambio* **2018**, *47*, 635–649. [CrossRef] [PubMed]
48. Fernández, D.S.; Lutz, M.A. Urban flash flood hazard zoning in Tucumán Province, Argentina, using GIS and multicriteria decision analysis. *Eng. Geol.* **2010**, *111*, 90–98. [CrossRef]
49. Dahri, N.; Abida, H. Monte carlo simulation-aided Analytical Hierarchy Process (AHP) for flash flood susceptibility mapping in Gabes Basin (Southeastern Tunisia). *Environ. Earth Sci.* **2017**, *76*, 302. [CrossRef]
50. Tehrany, M.S.; Pradhan, B.; Jebur, M.N. Spatial prediction of flash flood susceptible areas using rule based Decision Tree (DT) and a novel ensemble bivariate and multivariate statistical models in GIS. *J. Hydrol.* **2013**, *504*, 69–79. [CrossRef]
51. Li, K.; Wu, S.; Dai, E.; Xu, Z. Flash flood loss analysis and quantitative risk assessment in China. *Nat. Hazards* **2012**, *63*, 737–760. [CrossRef]
52. Garcia-Ruiz, J.M.; Regüés, D.; Alvera, B.; Lana-Renault, N.; Serrano-Muela, P.; Nadal-Romero, E.; Navas, A.; Latron, J.; Marti-Bono, C.; Arnáez, J. Flash flood generation and sediment transport in experimental catchments affected by land use changes in the central pyrenees. *J. Hydrol.* **2008**, *356*, 245–260. [CrossRef]
53. Benito, G.; Rico, M.; Sánchez-Moya, Y.; Sopeña, A.; Thorndycraft, V.R.; Barriendos, M. The impact of late holocene climatic variability and land use change on the flash flood hydrology of the Guadalentin River, Southeast Spain. *Glob. Planet. Chang.* **2010**, *70*, 53–63. [CrossRef]
54. Xu, Y.; Chung, S.-L.; Jahn, B.; Wu, G. Petrologic and geochemical constraints on the petrogenesis of Permian-Triassic Emeishan flash flood basalts in Southwestern China. *Lithos* **2001**, *58*, 145–168. [CrossRef]
55. Kazakis, N.; Kougias, I.; Patsialis, T. Assessment of flash flood hazard areas at a regional scale using an index-based approach and analytical hierarchy process: Application in Rhodope-Evros Region, Greece. *Sci. Total Environ.* **2015**, *538*, 555–563. [CrossRef] [PubMed]
56. Rahmati, O.; Pourghasemi, H.R.; Zeinivand, H. Flood susceptibility mapping using frequency ratio and weights-of-evidence models in the Golastan Province, Iran. *Geocarto Int.* **2016**, *31*, 42–70. [CrossRef]
57. Hall, M.A. Correlation-based feature selection of discrete and numeric class machine learning. In Proceedings of the Seventeenth International Conference on Machine Learning (ICML 2000), Stanford, CA, USA, 29 June–2 July 2000.
58. Pham, B.T.; Pradhan, B.; Bui, D.T.; Prakash, I.; Dholakia, M.B. A comparative study of different machine learning methods for landslide susceptibility assessment: A case study of Uttarakhand area (India). *Environ. Model. Softw.* **2016**, *84*, 240–250. [CrossRef]
59. Duma, M.; Twala, B.; Nelwamondo, F.V.; Marwala, T. Partial imputation to improve predictive modelling in insurance risk classification using a hybrid positive selection algorithm and correlation-based feature selection. *Curr. Sci.* **2012**, *103*, 697–705.
60. Freund, Y.; Schapire, R.E. A decision-theoretic generalization of on-line learning and an application to boosting. *J. Comput. Syst. Sci.* **1997**, *55*, 119–139. [CrossRef]
61. Pham, B.T.; Bui, D.T.; Prakash, I.; Dholakia, M.B. Hybrid integration of multilayer perceptron neural networks and machine learning ensembles for landslide susceptibility assessment at Himalayan Area (India) using GIS. *Catena* **2017**, *149*, 52–63. [CrossRef]
62. Breiman, L. Bagging predictors. *Mach. Learn.* **1996**, *24*, 123–140. [CrossRef]
63. Piao, Y.; Piao, M.; Jin, C.H.; Shon, H.S.; Chung, J.-M.; Hwang, B.; Ryu, K.H. A new ensemble method with feature space partitioning for high-dimensional data classification. *Math. Probl. Eng.* **2015**, *2015*, 1–12. [CrossRef]
64. He, Q.; Xu, Z.; Li, S.; Li, R.; Zhang, S.; Wang, N.; Pham, B.T.; Chen, W. Novel entropy and rotation forest-based credal decision tree classifier for landslide susceptibility modeling. *Entropy* **2019**, *21*, 106. [CrossRef]
65. Khosravi, K.; Cooper, J.R.; Daggupati, P.; Pham, B.T.; Bui, D.T. Bedload transport rate prediction: Application of novel hybrid data mining techniques. *J. Hydrol.* **2020**, 124774. [CrossRef]

66. Ting, K.M.; Witten, I.H. Stacking bagged and dagged models. In Proceedings of the 14th International Conference on Machine Learning, San Francisco, CA, USA, 8–12 July 1997.
67. Onan, A.; Korukouglu, S.; Bulut, H. Ensemble of keyword extraction methods and classifiers in text classification. *Expert Syst. Appl.* **2016**, *57*, 232–247. [CrossRef]
68. Thai, B.; Dieu, P.; Bui, T.; Prakash, I. Landslide susceptibility assessment using bagging ensemble based alternating decision trees, logistic regression and J48 decision trees methods: A comparative study. *Geotech. Geol. Eng.* **2017**, *35*, 2597–2611. [CrossRef]
69. Webb, G.I. Multiboosting: A technique for combining boosting and wagging. *Mach. Learn.* **2000**, *40*, 159–196. [CrossRef]
70. Kotti, M.; Benetos, E.; Kotropoulos, C.; Pitas, I. A neural network approach to audio-assisted movie dialogue detection. *Neurocomputing* **2007**, *71*, 157–166. [CrossRef]
71. Bui, D.T.; Ho, T.-C.; Pradhan, B.; Pham, B.-T.; Nhu, V.-H.; Revhaug, I. GIS-based modeling of rainfall-induced landslides using data mining-based functional trees classifier with adaboost, bagging, and multiboost ensemble frameworks. *Environ. Earth Sci.* **2016**, *75*, 1101.
72. Abellán, J.; Moral, S. Building classification trees using the total uncertainty criterion. *Int. J. Intell. Syst.* **2003**, *18*, 1215–1225. [CrossRef]
73. Mantas, C.J.; Abellán, J. Credal-C4.5: Decision tree based on imprecise probabilities to classify noisy data. *Expert Syst. Appl.* **2014**, *41*, 4625–4637. [CrossRef]
74. Abellán, J.; Masegosa, A.R. Combining decision trees based on imprecise probabilities and uncertainty measures. In *European Conference on Symbolic and Quantitative Approaches to Reasoning and Uncertainty*; Springer: Berlin/Heidelberg, Germany, 2007; pp. 512–523.
75. Dempster, A.P. Upper and lower probabilities induced by a multivalued mapping. In *Classic Works of the Dempster-Shafer Theory of Belief Functions*; Springer: New York, NY, USA, 2008; pp. 57–72.
76. Shafer, G. *A Mathematical Theory of Evidence*; Princeton University Press: Princeton, NJ, USA, 1976; p. 42.
77. Abellan, J.; Moral, S. Completing a total uncertainty measure in the dempster-shafer theory. *Int. J. Gen. Syst.* **1999**, *28*, 299–314. [CrossRef]
78. Abellan, J.; Moral, S. A non-specificity measure for convex sets of probability distributions. *Int. J. Uncertain. Fuzziness Knowl. Based Syst.* **2000**, *8*, 357–367. [CrossRef]
79. Mantas, C.J.; Abellán, J.; Castellano, J.G. Analysis of credal-C4. 5 for classification in noisy domains. *Expert Syst. Appl.* **2016**, *61*, 314–326. [CrossRef]
80. Walley, P. Inferences from multinomial data: Learning about a bag of marbles. *J. R. Stat. Soc. Ser. B* **1996**, *58*, 3–34. [CrossRef]
81. Mantas, C.J.; Abellán, J. Analysis and extension of decision trees based on imprecise probabilities: Application on noisy data. *Expert Syst. Appl.* **2014**, *41*, 2514–2525. [CrossRef]
82. Hong, H.; Panahi, M.; Shirzadi, A.; Ma, T.; Liu, J.; Zhu, A.-X.; Chen, W.; Kougias, I.; Kazakis, N. Flash flood susceptibility assessment in hengfeng area coupling adaptive neuro-fuzzy inference system with genetic algorithm and differential evolution. *Sci. Total Environ.* **2018**, *621*, 1124–1141. [CrossRef] [PubMed]
83. Pham, B.T.; Bui, D.T.; Dholakia, M.B.; Prakash, I.; Pham, H.V. A comparative study of least square support vector machines and multiclass alternating decision trees for spatial prediction of rainfall-induced landslides in a tropical cyclones area. *Geotech. Geol. Eng.* **2016**, *34*, 1807–1824. [CrossRef]
84. Ayalew, L.; Yamagishi, H.; Ugawa, N. Landslide susceptibility mapping using GIS-based weighted linear combination, the case in Tsugawa Area of Agano River, Niigata Prefecture, Japan. *Landslides* **2004**, *1*, 73–81. [CrossRef]
85. Van Dao, D.; Jaafari, A.; Bayat, M.; Mafi-Gholami, D.; Qi, C.; Moayedi, H.; Van Phong, T.; Ly, H.-B.; Le, T.-T.; Trinh, P.T. A spatially explicit deep learning neural network model for the prediction of landslide susceptibility. *Catena* **2020**, *188*, 104451.
86. Termeh, S.V.R.; Khosravi, K.; Sartaj, M.; Keesstra, S.D.; Tsai, F.T.C.; Dijksma, R.; Pham, B.T. Optimization of an adaptive neuro-fuzzy inference system for groundwater potential mapping. *Hydrogeol. J.* **2019**, *27*, 2511–2534. [CrossRef]
87. Pham, B.T.; Jaafari, A.; Prakash, I.; Singh, S.K.; Quoc, N.K.; Bui, D.T. Hybrid computational intelligence models for groundwater potential mapping. *Catena* **2019**, *182*, 104101. [CrossRef]

88. Tien Bui, D.; Shirzadi, A.; Chapi, K.; Shahabi, H.; Pradhan, B.; Pham, B.T.; Singh, V.P.; Chen, W.; Khosravi, K.; Ahmad, B.B.; et al. A hybrid computational intelligence approach to groundwater spring potential mapping. *Water* **2019**, *11*, 2013. [CrossRef]
89. Phong, T.V.; Phan, T.T.; Prakash, I.; Singh, S.K.; Shirzadi, A.; Chapi, K.; Ly, H.B.; Ho, L.S.; Quoc, N.K.; Pham, B.T. Landslide susceptibility modeling using different artificial intelligence methods: A case study at Muong Lay district, Vietnam. *Geocarto Int.* **2019**. [CrossRef]
90. Tien Bui, D.; Shirzadi, A.; Shahabi, H.; Geertsema, M.; Omidvar, E.; Clague, J.J.; Thai Pham, B.; Dou, J.; Talebpoor, D.; Lee, S.; et al. New ensemble models for shallow landslide susceptibility modeling in a semi-arid watershed. *Forests* **2019**, *10*, 743. [CrossRef]
91. Pham, B.T.; Prakash, I.; Singh, S.K.; Shirzadi, A.; Shahabi, H.; Bui, D.T. Landslide susceptibility modeling using Reduced Error Pruning Trees and different ensemble techniques: Hybrid machine learning approaches. *Catena* **2019**, *175*, 203–218. [CrossRef]
92. Pham, B.T.; Bui, D.T.; Pham, H.V.; Le, H.Q.; Prakash, I.; Dholakia, M.B. Landslide hazard assessment using random subspace fuzzy rules based classifier ensemble and probability analysis of rainfall data: A case study at Mu Cang Chai District, Yen Bai Province (Viet Nam). *J. Indian Soc. Remote Sens.* **2017**, *45*, 673–683. [CrossRef]
93. Jaafari, A.; Zenner, E.K.; Pham, B.T. Wildfire spatial pattern analysis in the Zagros Mountains, Iran: A comparative study of decision tree based classifiers. *Ecol. Inform.* **2018**, *43*, 200–211. [CrossRef]
94. Khosravi, K.; Shahabi, H.; Pham, B.T.; Adamowski, J.; Shirzadi, A.; Pradhan, B.; Dou, J.; Ly, H.; Grof, G.; Ho, H.L.; et al. A comparative assessment of flood susceptibility modeling using multi-criteria decision making analysis and machine learning methods. *J. Hydrol.* **2019**, *573*, 311–323. [CrossRef]
95. Khosravi, K.; Sartaj, M.; Tsai, F.T.; Singh, V.P.; Kazakis, N.; Melesse, A.M.; Prakash, I.; Bui, D.T.; Pham, B.T. A comparison study of drastic methods with various objective methods for groundwater vulnerability assessment. *Sci. Total Environ.* **2018**, *642*, 1032–1049. [CrossRef]
96. Miraki, S.; Zanganeh, S.H.; Chapi, K.; Singh, V.P.; Shirzadi, A.; Shahabi, H.; Pham, B.T. Mapping groundwater potential using a novel hybrid intelligence approach. *Water Resour. Manag.* **2019**, *33*, 281–302. [CrossRef]
97. Dou, J.; Yunus, A.P.; Bui, D.T.; Merghadi, A.; Sahana, M.; Zhu, Z.; Chen, C.; Khosravi, K.; Yang, Y.; Pham, B.T. Assessment of advanced random forest and decision tree algorithms for modeling rainfall-induced landslide susceptibility in the Izu-Oshima Volcanic Island, Japan. *Sci. Total Environ.* **2019**, *662*, 332–346. [CrossRef]
98. Abedini, M.; Ghasemian, B.; Shirzadi, A.; Shahabi, H.; Chapi, K.; Pham, B.T.; Ahmad, B.B.; Tien Bui, D. A novel hybrid approach of bayesian logistic regression and its ensembles for landslide susceptibility assessment. *Geocarto Int.* **2018**. [CrossRef]
99. Chang, K.T.; Merghadi, A.; Yunus, A.P.; Pham, B.T.; Dou, J. Evaluating scale effects of topographic variables in landslide susceptibility models using GIS-based machine learning techniques. *Sci. Rep.* **2019**, *9*, 1–21. [CrossRef] [PubMed]
100. Nohani, E.; Moharrami, M.; Sharafi, S.; Khosravi, K.; Pradhan, B.; Pham, B.T.; Lee, S.; Melesse, A.M. Landslide susceptibility mapping using different GIS-based bivariate models. *Water* **2019**, *11*, 1402. [CrossRef]
101. Pham, B.T.; Nguyen, M.D.; Bui, K.T.; Prakash, I.; Chapi, K.; Bui, D.T. A novel artificial intelligence approach based on multi-layer perceptron neural network and biogeography based optimization for predicting coefficient of consolidation of soil. *Catena* **2019**, *173*, 302–311. [CrossRef]
102. Nguyen, V.V.; Pham, B.T.; Vu, B.T.; Prakash, I.; Jha, S.; Shahabi, H.; Shirzadi, A.; Ba, D.N.; Kumar, R.; Chatterjee, J.M.; et al. Hybrid machine learning approaches for landslide susceptibility modelling. *Forests* **2019**, *10*, 157. [CrossRef]
103. Pham, B.T.; Prakash, I.; Jaafari, A.; Bui, D.T. Spatial prediction of rainfall-induced landslides using aggregating one-dependence estimators classifier. *J. Indian Soc. Remote Sens.* **2018**, *46*, 1457–1470. [CrossRef]
104. Pham, B.T.; Prakash, I. A novel hybrid model of bagging-based naïve bayes trees for landslide susceptibility. *Bull. Eng. Geol. Environ.* **2019**, *78*, 1911–1925. [CrossRef]
105. Pham, B.T. A novel classifier based on composite hyper-cubes on iterated random projections for assessment of landslide susceptibility. *J. Geol. Soc. India* **2018**, *91*, 355–362. [CrossRef]
106. Dou, J.; Yunus, A.P.; Xu, Y.; Zhu, Z.; Chen, C.W.; Sahana, M.; Yang, Y.; Khosravi, K.; Pham, B.T. Torrential rainfall-triggered shallow landslide characteristics and susceptibility assessment using ensemble data-driven models in the Dongjiang Reservoir Watershed, China. *Nat. Hazards* **2019**, *97*, 579–609. [CrossRef]

107. Pham, B.T.; Prakash, I.; Dou, J.; Singh, S.K.; Trinh, P.T.; Tran, H.T.; Le, T.M.; Phong, T.V.; Khoi, D.K.; Shirzadi, A.; et al. A novel hybrid approach of landslide susceptibility modelling using rotation forest ensemble and different base classifiers. *Geocarto Int.* **2019**, 1–25. [CrossRef]

108. Peng, Y.; Shi, Y.; Yan, H.; Chen, K.; Zhang, J. Coincidence risk analysis of floods using multivariate copulas: Case study of Jinsha River and Min River, China. *J. Hydrol. Eng.* **2018**, *24*, 05018030. [CrossRef]

109. Le, L.M.; Ly, H.B.; Pham, B.T.; Le, V.M.; Pham, T.A.; Nguyen, D.H.; Tran, X.T.; Le, T.T. Hybrid artificial intelligence approaches for predicting buckling damage of steel columns under axial compression. *Materials* **2019**, *12*, 1670. [CrossRef] [PubMed]

110. Ly, H.B.; Desceliers, C.; Le, L.M.; Le, T.T.; Pham, B.T.; Nguyen-Ngoc, L.; Doan, V.T.; Le, M. Quantification of uncertainties on the critical buckling load of columns under axial compression with uncertain random materials. *Materials* **2019**, *12*, 1828. [CrossRef] [PubMed]

111. Shahabi, H.; Jarihani, B.; Tavakkoli Piralilou, S.; Chittleborough, D.; Avand, M.; Ghorbanzadeh, O. A semi-automated object-based gully networks detection using different machine learning models: A case study of Bowen Catchment, Queensland, Australia. *Sensors* **2019**, *19*, 4893. [CrossRef] [PubMed]

112. Jalayer, F.; De Risi, R.; De Paola, F.; Giugni, M.; Manfredi, G.; Gasparini, P.; Topa, M.E.; Yonas, N.; Yeshitela, K.; Nebebe, A.; et al. Probabilistic GIS-based method for delineation of urban flooding risk hotspots. *Nat. Hazards* **2014**, *73*, 975–1001. [CrossRef]

113. Chapman, L. Increasing vulnerability to floods in new development areas: Evidence from Ho Chi Minh City. *Int. J. Clim. Chang. Strateg. Manag.* **2018**. [CrossRef]

114. Dano, U.L.; Balogun, A.L.; Matori, A.N.; Wan Yusouf, K.; Rimi Abubakar, I.; Mohamed, S.; Aina, Y.A.; Pradhan, B. Flood susceptibility mapping using GIS-based analytic network process: A case study of Perlis, Malaysia. *Water* **2019**, *11*, 615. [CrossRef]

115. Zhao, G.; Pang, B.; Xu, Z.; Peng, D.; Xu, L. Assessment of urban flood susceptibility using semi-supervised machine learning model. *Sci. Total Environ.* **2019**, *659*, 940–949. [CrossRef]

116. Khosravi, K.; Melesse, A.M.; Shahabi, H.; Shirzadi, A.; Chapi, K.; Hong, H. Flood susceptibility mapping at Ningdu catchment, China using bivariate and data mining techniques. In *Extreme Hydrology and Climate Variability*; Elsevier: London, UK, 2019; pp. 419–434.

117. Termeh, S.V.R.; Kornejady, A.; Pourghasemi, H.R.; Keesstra, S. Flood susceptibility mapping using novel ensembles of adaptive neuro fuzzy inference system and metaheuristic algorithms. *Sci. Total Environ.* **2018**, *615*, 438–451. [CrossRef]

118. Ahmadlou, M.; Karimi, M.; Alizadeh, S.; Shirzadi, A.; Parvinnejhad, D.; Shahabi, H.; Panahi, M. Flood susceptibility assessment using integration of adaptive network-based fuzzy inference system (ANFIS) and biogeography-based optimization (BBO) and BAT algorithms (BA). *Geocarto Int.* **2019**, *34*, 1252–1272. [CrossRef]

119. Thai Pham, B.; Tien Bui, D.; Prakash, I. Landslide susceptibility modelling using different advanced decision trees methods. *Civil Eng. Environ. Syst.* **2018**, *35*, 139–157. [CrossRef]

120. Li, H.; Ouyang, J.; Li, F.; Xie, X. Study on safety evaluation model of small and medium-sized earth-rock dam based on BP-AdaBoost algorithm. In *IOP Conference Series: Materials Science and Engineering*; IOP Publishing: Bristol, UK, 2019; p. 032024.

121. Avand, M.; Janizadeh, S.; Tien Bui, D.; Pham, V.H.; Ngo, P.T.T.; Nhu, V.H. A tree-based intelligence ensemble approach for spatial prediction of potential groundwater. *Int. J. Digital Earth* **2020**, 1–22. [CrossRef]

122. Kuncheva, L. *Combining Pattern Classifiers Methods and Algorithms*; John Wiley&Sons. Inc. Publication: Hoboken, NI, USA, 2014.

123. Thai Pham, B.; Shirzadi, A.; Shahabi, H.; Omidvar, E.; Singh, S.K.; Sahana, M.; Asl, D.T.; Ahmad, B.B.; Quoc, N.K.; Lee, S. Landslide susceptibility assessment by novel hybrid machine learning algorithms. *Sustainability* **2019**, *11*, 4386. [CrossRef]

124. Dou, J.; Yunus, A.P.; Bui, D.T.; Merghadi, A.; Sahana, M.; Zhu, Z.; Chen, C.; Han, Z.; Pham, B.T. Improved landslide assessment using support vector machine with bagging, boosting, and stacking ensemble machine learning framework in a mountainous watershed, Japan. *Landslides* **2019**. [CrossRef]

125. Merghadi, A.; Abderrahmane, B.; Tien Bui, D. Landslide susceptibility assessment at Mila Basin (Algeria): A comparative assessment of prediction capability of advanced machine learning methods. *ISPRS Int. J. Geo-Inf.* **2019**, *7*, 268. [CrossRef]

126. Gautam, D.; Dong, Y. Multi-hazard vulnerability of structures and lifelines due to the 2015 Gorkha earthquake and 2017 central Nepal flash flood. *J. Build. Eng.* **2018**, *17*, 196–201. [CrossRef]
127. Eem, S.-h.; Yang, B.-j.; Jeon, H. Simplified methodology for urban flood damage assessment at building scale using open data. *J. Coast. Res.* **2018**, *85*, 1396–1400. [CrossRef]

© 2020 by the authors. Licensee MDPI, Basel, Switzerland. This article is an open access article distributed under the terms and conditions of the Creative Commons Attribution (CC BY) license (http://creativecommons.org/licenses/by/4.0/).

Article

Modeling Urban Flood Inundation and Recession Impacted by Presence of Manholes

Merhawi GebreEgziabher * and Yonas Demissie

Department of Civil and Environmental Engineering, Washington State University, Richland, WA 99354, USA; y.demissie@wsu.edu
* Correspondence: me.gebremichael@wsu.edu

Received: 16 March 2020; Accepted: 14 April 2020; Published: 18 April 2020

Abstract: Urban flooding, caused by unusually intense rainfall and failure of storm water drainage, has become more frequent and severe in many cities around the world. Most of the earlier studies focused on overland flooding caused by intense rainfall, with little attention given to floods caused by failures of the drainage system. However, the drainage system contributions to flood vulnerability have increased over time as they aged and became inadequate to handle the design floods. Adaption of the drainages for such vulnerability requires a quantitative assessment of their contribution to flood levels and spatial extent during and after flooding events. Here, we couple the one-dimensional Storm Water Management Model (SWMM) to a new flood inundation and recession model (namely FIRM) to characterize the spatial extent and depth of manhole flooding and recession. The manhole overflow from the SWMM model and a fine-resolution elevation map are applied as inputs in FIRM to delineate the spatial extent and depth of flooding during and aftermath of a storm event. The model is tested for two manhole flooding events in the City of Edmonds in Washington, USA. Our two case studies show reasonable match between the observed and modeled flood spatial extents and highlight the importance of considering manholes in urban flood simulations.

Keywords: manhole flooding urban flooding; grid-based modeling; SWWM; FIRM

1. Introduction

Flooding is one of the most frequent weather-related natural disasters and affects many people around the world every year [1]. Major floods often cause significant impacts to communities and economies [1,2]. For example, in the United States alone, flood damages cost $260 billion (USD) per year from 1980 to 2013 [3]. The National Flood Insurance Program (NFIP) paid on average $2.9 billion a year between 2000 and 2018 [4]. Similarly, flooding caused more than 700 fatalities and at least €25 billion economic losses in Europe between 1998 and 2004 [5]. Global warming is expected to lead to more frequent extreme precipitation events, increasing flood hazards in many cities around the world [2,6,7].

Despite the severe damages that flooding can induce, floods are an important part of life in various regions of the world where people have been adapting for centuries. They support riparian ecosystems dependent on flood inundated zones [8], and they are the principal source of groundwater recharge in many arid and semi-arid settings [9–14]. The negative impacts of floods, however, are considerable, and are mostly associated with the unexpected magnitudes and frequencies of floods, which are connected to climate change, rapid expansion of urbanized areas, and inadequate and aging urban drainage systems [2,15–20].

Urban flooding and associated damages to properties account for 73% of the $107.8 billion total damages caused by floods from 1960 to 2016 in the United States [21]. Thus, accurate flood monitoring and estimation are essential to reduce flood impacts and vulnerabilities while supporting urban planning and ecosystems.

The spatial and temporal characteristics of floods in urban areas are complex due to the widespread change to the land uses [22], which introduces micro-urban features such as buildings, roads and drainage networks [23,24]. The specific urban infrastructures that affect flood form a storm event includes the type and geometry of building, garage ramps, light wells, pillars, and yards at or just beneath the ground surface [25,26]. Overall, it is an established concept that an increase in impervious areas and connected conveyance systems in urban areas increase peak discharges and volumes [27,28]. However, flood mitigation and management requires detailed information about the spatial extents of floods, their water levels (i.e., flood depth), and flow velocities [29]. Such information is often impractical to measure directly. Consequently, empirical and hydraulic models are widely used to estimate these parameters [29–31] and assess associated flooding risk [18,32–36].

Although hydrological modeling can simulate both surface and subsurface processes adequately at a watershed level, most of them are unable to simulate urban flooding accurately [37]. This is partly due to the difficulty of defining model boundary conditions and the complex nature of flood propagation in urban areas. In addition, most flood inundation models lack the potential contribution of manholes overflow to the flooding. The majority of these models are commercial (e.g., MIKE FLOOD, XPSWMM, and FLO2D), and are not accessible for most users. They are complex, require a large number of datasets, and computational resources, which make them ineffective for most applications [29,38]. Thus, simplified flood inundation modeling techniques are commonly used to determine the spatial extent of flooding in urban areas [37,39]. These models only require a digital elevation map and mathematical representation of floodwater propagation in a given area [40]. These models were used to identify the potential flood-prone regions during a given storm event, but not the aftermath of the flood event. Furthermore, the simplified models often do not simulate the watershed and the drainage system. Consequently, coupling hydrologic models, one-dimensional (1D) hydrodynamic models, and simplified flood inundation models are gaining attention as an alternative approach to simulate flood inundations in urban areas [29].

Despite well-documented effects of manholes on urban flooding [19,20,37,41,42], there exists relatively limited research [43–47] that directly incorporate manhole overflow in simulations of urban flooding. Chen et al. [48] used coupled surface and sewer flow modeling and found approximately 2 m surge flow from manholes. Leandro and Martins [43] used a two-dimensional (2D) flood inundation model to simulate bi-directional flow interaction between sewer and overland flow, to estimate the volume of sewer serge from manholes ranging from 15,992–18,404 m^3 and a maximum possible depth of 0.8 m. Son et al. [44] coupled a one-dimensional storm water management model with a 2D overland flow model from a manhole overland flow with a maximum overland flow, ranging from 2–5 m^3/s at different manhole locations, and estimated a 0.9 m flood depth and 2.5 m/s flood velocity. Jang et al. [46] used a coupled one-dimensional sewer flow with the two-dimensional overland flow and estimated manhole overflow depth up to 2 m. Seyoum et al. [47] coupled 1D sewer and 2D dimensional flood inundation models to simulate urban flooding and showed the combined flood depth variations from 0.3 to 0.8 m in their study area. Manhole overflow is a critical issue in urban areas, yet their contribution to flooding is not well understood. Most hydrodynamic models assume that excess water will pond around the manhole and return back or will be lost from the system after the flood recedes [49]. Major cities have aging infrastructures and drainage systems designed and built more than a decade years ago [50]. These systems are increasingly underperforming due to their design assumption of stationary storms and flood events [51,52]. This assumption often causes inadequacy to handle the rising flood risk caused by increased storms and impervious layers.

The main objective of the study is to develop and test a new flood inundation and recession method (FIRM) that can readily be used to simulate excess floodwaters generated from manholes. The United States Environmental Protection Agency (EPA)-Storm Water Management Model (SWMM) is commonly used to simulate the complex hydrological and hydraulic processes in urban areas, but it does not have the configuration to simulate flood inundation and recession from manholes overflows. Our method uses manhole overland flow volumes and depths from SWMM output and digital elevation data of the study area to simulate flood inundation, recession, and depth. The flood inundation model uses the flat-water assumption to distribute the overflow. The recession model uses the location of manhole and surrounding topographic variation to determine whether the inundated region is going to drain to the manhole or pond in localized regions. The model was tested using a synthetic case study and a flooding event in Edmonds, an urban region in Washington State. Since the inputs (elevation and overflow) and outputs (flood area and depth) are known, the synthetic case study was used as a proof-of-concept to validate the model accuracy before applying it to a real-world problem. It allowed us to test the model performance using different scenarios under variable hypothetical case studies. For our Edmonds case study, previous study reports, social media, and news reports were used to delineate the flood boundary (i.e., used as observational data). The flooded area estimated by FIRM was compared against the reconstructed flood area visually and using statistical measures to evaluate our model's ability to identify areas that were flooded versus dry during the actual flood event.

2. Methods

This section details (a) the hydrodynamic model "SWMM" applied to Edmond case study (Section 2.1), (b) the model domain, calibration, and validation (Section 2.2), (c) the simulation of manholes' overflow and recession using FIRM (Section 2.3), and (d) the evaluation of the FIRM performance using statistical measures (Section 2.4).

In addition to the synthetic case study used for proof-of-concept, the FIRM was applied to simulate manhole flooding in the city of Edmonds. The SWMM model domain was discretized into detailed sub-catchments to incorporate key hydrologic and hydraulic components. Three years of sub-daily rainfall and lake level data were used to develop and calibrate the model using the automated differential evolution optimization methods [53]. The SWMM model was validated against a one-year simulation. Following the SWMM simulation, we simulated the spatial extent and depth of flood inundation and recession from flooded manholes. A high-resolution digital surface model (DSM) was used due to its detail information of surface features, including infrastructures. Unlike to the digital elevation map (DEM), the DSM contains surface elevation and surface features [54].

2.1. Hydrodynamic Modeling Using SWMM

SWMM is one of the most widely used hydrologic and hydraulic models in urban settings [55]. It is capable of simulating event-based or continuous rainfall-runoff processes that are useful for both water quality and quantity analyses in urban areas [56].

SWMM uses spatially distributed and temporally discrete processes to simulate the hydrological and hydraulic state variables [49]. As the simulation progresses, the state variables will be updated and stored as follow.

$$X_t = f(X_{t-1}, I_t, P), \tag{1}$$

$$Y_t = g(X_t, P), \tag{2}$$

where f and g represent the functions that calculate and update the state and output variables respectively. X_t represents state variables (such as flow rate and depth in a drainage network link), Y_t represents output variables (such as runoff flow rate at each sub-catchments and outlet), P represents the constant parameter, I_t represents input variables (such as rainfall and temperature) at a given time.

The hydraulic simulation of SWMM involves water and contaminants transport through the conveyance portion of the drainage network. External flow sources entered the drainage network using inflow nodes and transported through pipes and storage components and finally exit at outflow nodes [49]. The flow equation through links can be solved using the dynamic and kinematic wave routing equation. The dynamic wave analysis uses the Saint-Venant flow equation, whereas the kinematic wave method uses the simplified form of the momentum equation to estimate the flow condition though the conveyance system [49]. The dynamic wave analysis has advantages in simulating gradually varied flow conditions, such as surge, in the drainage systems. This results in having the dynamic wave method, to have different pressure and friction force part of the momentum. We used a dynamic wave routing method to simulate the gradually varied flow in urban drainage system.

2.2. Manhole Overland Flow Inundation and Recession Modeling

SWMM considers overflow from manholes as either surface ponding for a certain period and return back to the drainage or lose from the system [49]. However, these approaches do not allow for direct estimation of the associated flood spatial and temporal extents. Besides, all the overland flow generated from the node does not necessarily return to the drainage, as some will recess, and others might be isolated from the manholes and ponded in depressions. In this study, we developed a simplified grid-based flood module to propagate and recess overflow from and to a manhole spatially. The module requires a gridded surface elevation map, locations of manholes, and total overflow volume and depth at the manholes. The flood inundation computation assumes that (i) water flows from a higher elevation to lower elevation because of gravity, (ii) water spreads spatially by maintaining its level surface ('flat' water assumption), and (iii) flood fills first the nearest and connected cell with the lowest elevation. The module estimates both the flood depth and spatial extent by propagating the excess pressurized water from the manholes to the surrounding areas according to the topographic variations. A high-resolution (1 m × 1 m) digital surface terrain from Light Detection and Ranging (LiDAR) was used to capture the flooding extents accurately.

The flood recession module uses the location of manholes, the areal extent of the flood, as well as topography of the region to determine the aftermath of the flooded area. The flood recession method assumes that all flooded cells drain back to the manholes if there exists a flow path connecting them to the manholes; otherwise, it will remain ponded. Some of the flooded water in the local depressions can be disconnected from the manhole and will not fully be drained. Otherwise, the ponded locations can be distant from the manholes due to the topographic barrier.

2.2.1. Manhole Overland Flow Inundation Modeling

Our simulation of flood inundation is based on the elevation variations in neighboring cells. Starting from the cells containing the flooded manhole, floodwaters propagate to neighboring cells if the cell elevation is lower-than and connected-to a flooded cell. We used the D8 neighboring algorithm, which evaluates the elevations of the eight adjacent neighboring cells to each flooded cell, to determine the preferred flow direction (Figure 1). The flooded cells will maintain level flood surface and expand spatially as the lowest elevations are filled with available excess water. For multiple manholes overland flow, iterations are used to distribute all the overflows by maintaining the level flooded surface.

Before the flooding starts, the area is assumed dry, or the flood depths at each grid cell are known. Once the manhole overflow starts, the floodwater inundates neighboring and connected grid cells that have lower elevations (Figure 1). This inundation process will stop when the excess overland flow is completely allocated, and no water left to inundate the next dry cell. The flood inundation calculation is performed based on three main cases. In the first case, if the manhole cell elevation is smaller than that of the neighboring cells, the overflow will accumulate in the manhole cell until it reaches the minimum elevation of the neighboring grid cell. Once the overflow surface elevation exceeds the minimum elevation of a neighboring cell, the water will propagate spatially as long as there is enough overflow. In the second case, if the elevation of the manhole is higher than the elevations of neighbor cells, the overflow will be allocated automatically to the neighboring cells regardless of their slopes but maintaining level flooded surface. In the third case, if the elevation of the manhole is the same as the elevation of any neighboring cell, the excess water will be distributed equally among cells with the same elevations.

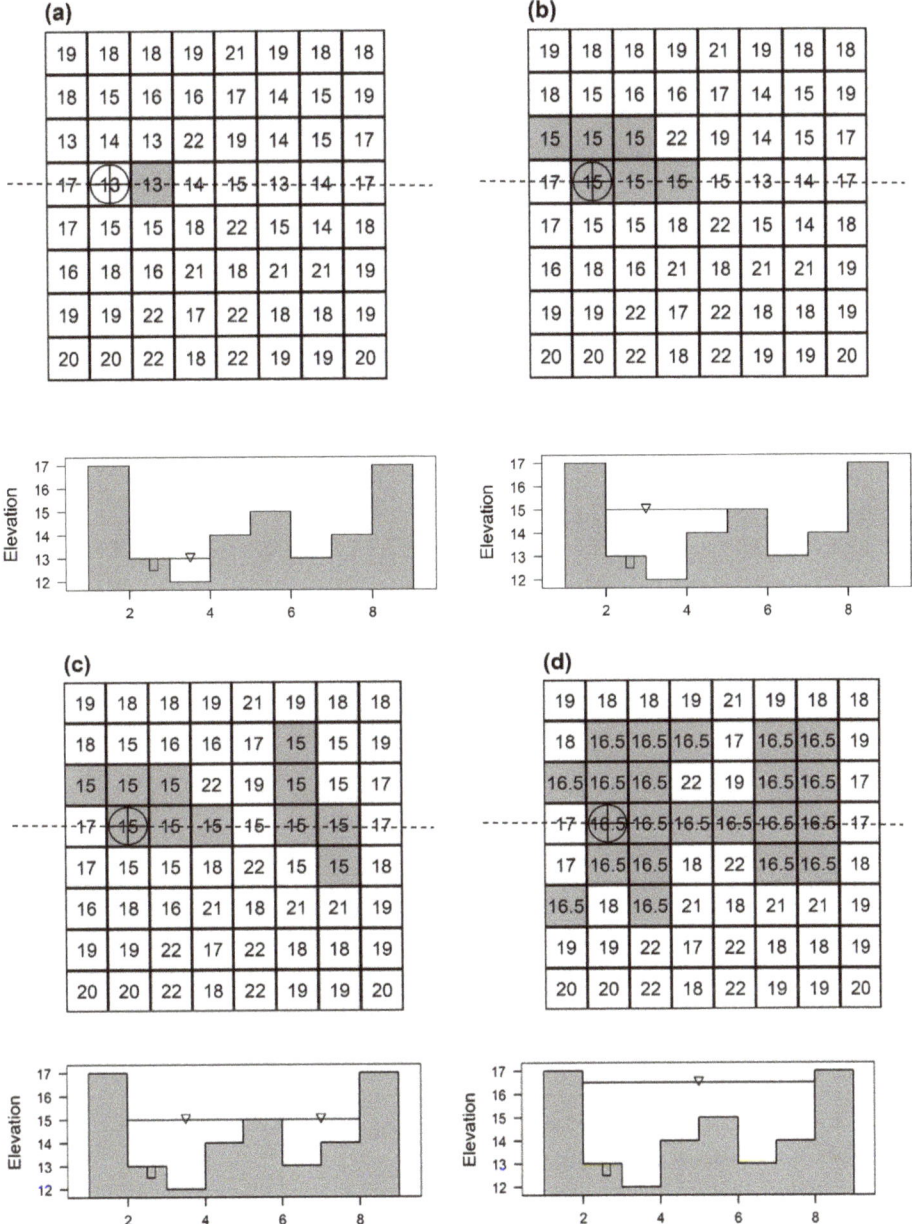

Figure 1. Schematic representation of the flood inundation modeling from one-manhole (the circle with cross) for different amounts of manhole overflow: (**a**) 1 unit, (**b**) 11 units, (**c**) 17 units, and (**d**) 46 units. The numbers inside the unshaded and unshaded grids represent surface elevation and flood levels, respectively. Dashed line represents the profile line.

Figures 1 and 2 show a schematic representation of flood inundations caused by overflow from one manhole and two manholes respectively. When the flood surface (surface elevation plus inundation depth) exceeds the elevation of the surrounding cells, the water will start to flow to the cells with the lowest elevations. As the overland flow increases, the neighboring cells will be gradually flooded while maintaining level flood surface. For example, when the overland volume is 49 units (Figure 1d), the extent of the flood coverage is first determined, based on the amount of the overland flow and elevation, and then the floodwater is distributed iteratively among the grids within the flooded area to maintain level flood surface. This iterative process considers the available flood volume, and recursively fills the next small neighboring cells.

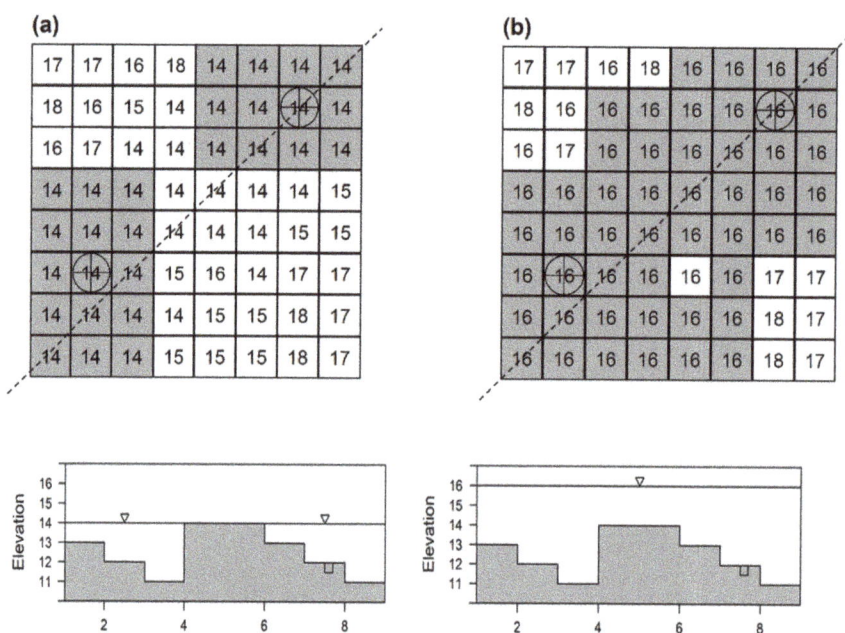

Figure 2. Schematic diagram showing flood inundation from two manholes (the circles with cross) for different amounts of manhole overflow: (**a**) 22 units from the left manhole and 19 units from the right manhole, (**b**) 110 units from the left manhole and 111 units from the right manhole. The numbers inside the unshaded and unshaded grids represent surface elevation and flood levels, respectively. Dash line represents the profile line.

Not explicitly considering the slope and land covers are the main limitation of the presented flood inundation approach. Similar to other grid-based hydrological and hydraulic modeling, the DSM resolution poses a high level of uncertainty in the model representation. Thus, the use of fine-resolution DSM and surface maps are critical for estimating the floods from manholes accurately [37]. For this study, we have used a one-meter resolution Light Detection and Ranging (LiDAR) DSM obtained from Washington State Department of Natural Resources. The model and the test cases do not include manhole inundation caused my previous storm and overland flow the drainage. Thus, we assumed dry terrain prior to the overflow from the manholes. However, in most cases, as the reviewer correctly stated, the manholes and their surrounding areas might already be flooded before the overflow. One simple modification is to adjust the terrain elevations for the inundation depth and then simulate the manhole overflow on top. Such an approach may work if the inundations surrounding the manholes that resulted from the previous storm were stationary, and have known flood depths. However, in most cases, the overland and overflow inundations from the drainage area and manholes, respectively, happen simultaneously, requiring a coupled simulation of both inundations.

2.2.2. Recession Modeling Associated with Manholes

Similar to the overland flow inundation, the flood recession to a given manhole considers the elevations of the surrounding grid cells and their connectivity to the manhole. For any given cell, the flooded water above the elevations of the adjacent grid cells will drain to the manhole if a flow path exists. Otherwise, it will remain ponded in the depressions. For example, when a manhole elevation is higher than the neighboring cells, only partial recession will occur. Starting from the manhole cell and expanding outward using the D8 algorithm, the recession model identifies flooded cells and drain them completely if their elevations are above one of the eight neighboring cells. Else drains part of the flooded water above the minimum elevation of the eight neighboring cells. As the grid cell expand spatially by searching adjacent eight neighboring, the connectivity of the target cell and the neighbor is traced using a flow path algorithm. For each flooded cell, the flow path algorithm detects the presence of any possible flow path to the manhole.

Figure 3 shows examples of flood recession based on single and multiple manholes. The result from the flood inundation estimate is given in Figure 3a. It is assumed that when rainfall ceased, the holding capacity of the drainage system decreases, and overland flow starts to recess toward the manholes. Figure 3b,c illustrate how flooded surfaces are recessing for a single and two manholes scenarios. The recession associated with a single manhole (Figure 3b) indicates two regions of ponding—one next to the manhole and another away from manhole caused by local topographic barrier. The ponded regions away from the manhole are not connected to the manhole hydraulically. The second case introduced an additional manhole in the depression region, and thereby the flood can be drained by the two independent manholes. The left manhole recess similarly as Figure 3b. The second manhole drains the ponded floodwater caused by a topographic barrier (Figure 3c).

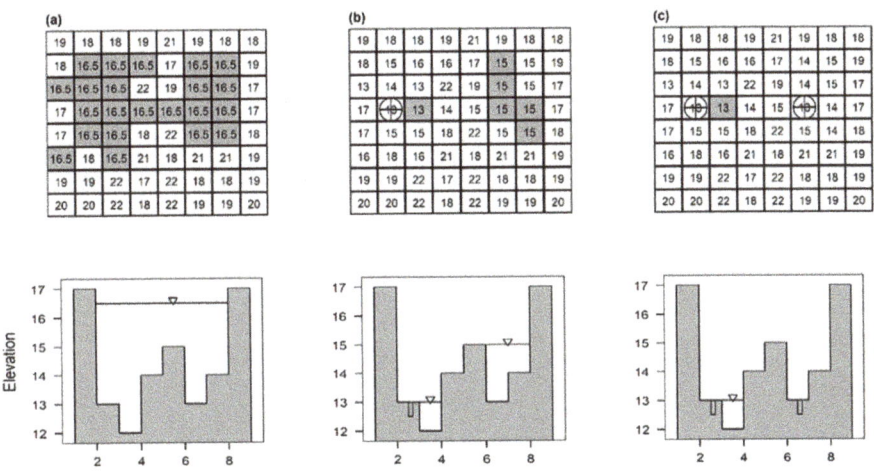

Figure 3. Schematic representation of flood recession processes from a single and multiple manhole. (**a**) Represent the areal extent and profile section of the flooded regions, (**b**) represents the areal extent and profile of flood recession for a single manhole, and (**c**) show a recession surface and profile after drained by two manholes.

2.3. Study Site: The Hall Creek Watershed

The Hall Creek watershed is an urban watershed (Figure 4), containing four major cities near Seattle, WA. These cities are Edmonds, Esperance, Lynnwood, and Mountlake Terrace, which are part of the northern Seattle–Tacoma–Bellevue metropolitan region. The predominant land cover type in the watershed is a developed urban region, which accounts for 96% of the land cover. The rest is covered by forest and water bodies (Figure 4b). The watershed is frequently affected by prolonged storm and flooding events.

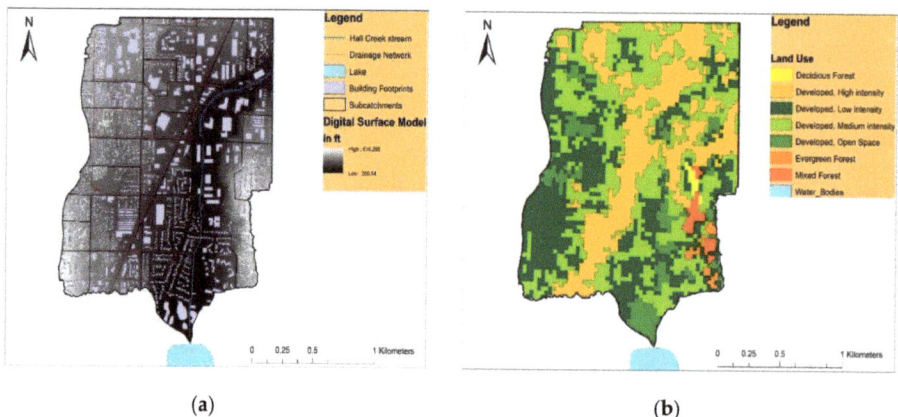

(a)   (b)

Figure 4. Hall Creek watershed showing (**a**) buildings, drainage network, and the surface elevation as a background, and (**b**) land cover map.

The Hall Creek is intermittent and drains toward Lake Ballinger. It is the main tributary to the Lake Ballinger [57,58], which discharges to the downstream McAleer Creek. The creek does not have a monitoring discharge and stage data. Since the lake level is highly influenced by storm events (Figure 5) and the flow from the Hall Creek, the available level data were used to calibrate the SWMM model.

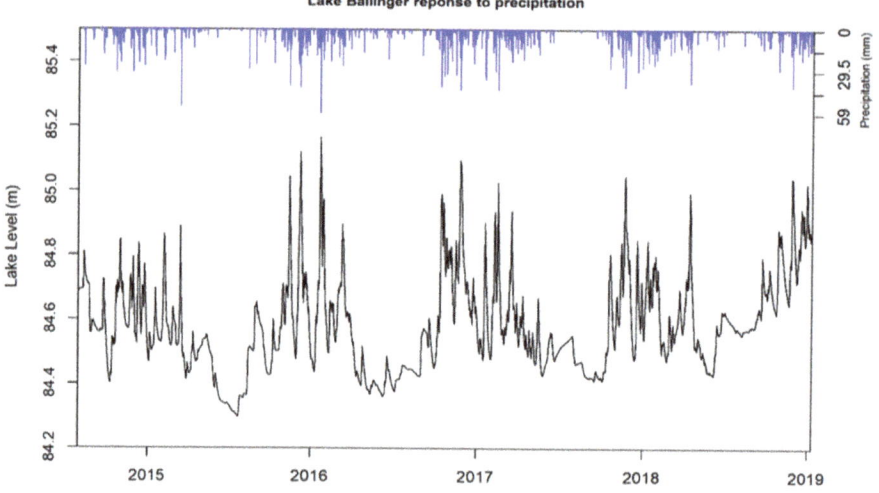

Figure 5. Time series of the Lake Ballinger level and precipitation in the watershed.

The flooding in the study area is mostly caused by storm events, while the urban area and street intersections are often affected by manhole overflow. For example, during the 19 September 2016 storm events, two manholes overflowed and caused flooding in the surrounding area. This flood event was used to validate our flood inundation and recession methodologies. One of the manholes flooded areas across a highway (Case 1), while the other caused flooding alongside the highway (Case 2) (Figure 6).

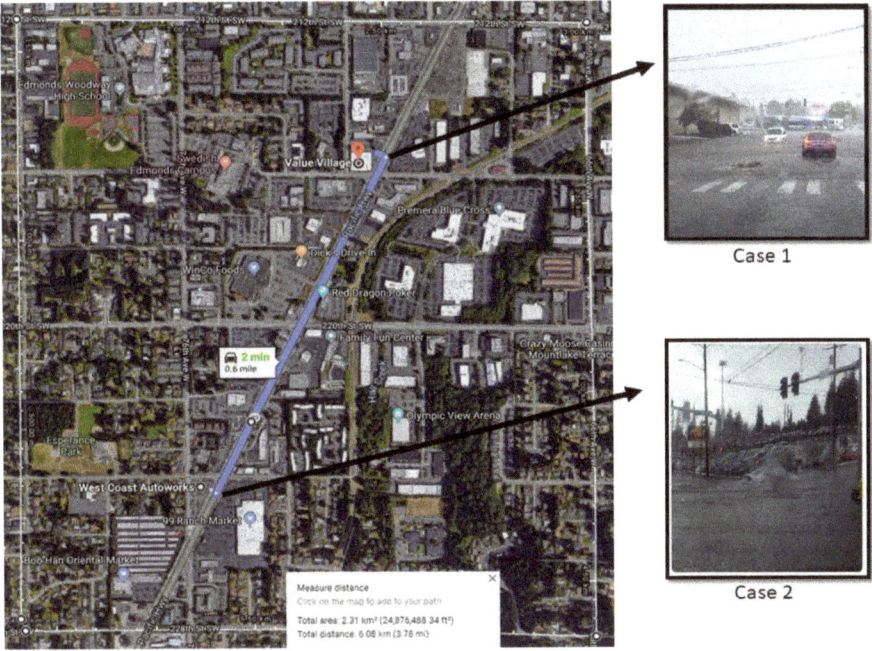

Figure 6. The location of the two manholes, which cause flood along a street (Case 2) and across a street (Case 1).

2.3.1. Data

High resolution observed meteorological data (obtained from the King County's watersheds and rivers database) and sub-hourly lake level data (obtained from the city of Edmonds) were used to develop the SWMM model. Surface elevation from a one-meter resolution LiDAR data (obtained from Washington State Department of Natural Resources) were used for the flood inundation and recession modeling. We have extracted the conveyance network system of the city of Edmonds and Mountlake Terrace from their respective Geographical Information System (GIS) Department. We have identified the type, location, and possible flow direction of the sewer system. The land cover data were obtained from the US Department of Agriculture and were used to estimate the percentage of impervious layer. For the purpose of calibration and validation of the FIRM model, the observed flood boundaries are delineated using related images (Figure 6) and texts from social media users (such as locations). We used Google Earth and high-resolution LiDAR data to determine the relative topographical variation and delineated the inferred flood boundary by considering 360-street and panoramic view. We also used the city of Edmonds GIS-dataset to correlate the inferred flood boundary and building footprints to check if there exists a mismatch between the building blocks and the street boundaries.

2.3.2. SWMM Model

The SWMM model was discretized into 32 sub-catchments based on the hydrological and drainage network criteria. These criteria include percent of land cover, slope, availability of conveyance network system, and percent of impervious layers. A 5% of land use, slope, and soil type is used to subdivide the catchment. The sub-catchment layer, conduit layer and node layers are used for urban watershed discretization. After the watershed was discretized, the external inputs such as precipitation, temperature, and evapotranspiration were extracted and applied for each sub-catchment. The hydraulic properties of manholes, storage, ditches, culverts, and other structures were incorporated. There are total of 106 manholes, one storage, and 108 nodes connected by conduits in the study area. Only 40% of the manholes were considered in our study based on consideration of data availabilities and the computational requirement of the SWMM model. The Hall Creek flux are represented using SWMM's inflow package. Despite the reasonable performance of our model in representing the hydrology and hydraulics of the study area (Section 3.1), not considering all the hydraulic structures in our model might have introduced some level of uncertainty that need for a further study. SWMM simulations can take hours if the model domain is large and accompanied by detailed complex hydraulic structures and sub-daily meteorological and hydrological input variables. The model simulation was conducted using the "swmmr" R-package [55].

Model calibration can be performed using either manual or automated method [59]. In this study, we used both methods of calibration to ensure usage of their advantages. First, we used manual calibration to identify sensitive parameters. The sensitive parameters was then further calibrated using differential evolution (DE) method, which finds the global optimum parameters values for continuous and differentiable functions [60], based on successive generation and transformation of the parameters values under a given fitness-measured criteria [53]. The DE requires defining parameters upper and lower bounds, objective function, and the lower or the upper optimal solution goal. The algorithm starts by randomly dividing the parameters values in to three distinct populations. The parameters values from each population are then combined to generate the next sets of populations that minimize the objective function. To ensure global optimal solution, the algorithm uses mutation to include non-optimal parameters values in the new populations [53]. The evolution continues until it meets the objective function criteria.

Due to the lack of discharge observational data for the Hall Creek, the model calibration and validation were performed based on the lake level fluctuation of Lake Ballinger, which is located at the outlet of the creek. Since the creek is a main feed to the lake, the lake level fluctuations reflect the changes in the creek discharge. The model calibration includes the initial condition of the model, which enables to determine the calibration parameter ranges, and optimization of the parameters using differential evolution optimized method [61]. Nearly three (2.7) years of data were used to calibrate the model, and one-year of data were used for validation.

The initialization was used to identify sensitive parameters and their respective parameter ranges. The model was then optimized using the differential evolution method, namely the "DEoptim" packages [61]. Differential Evolution (DE) is a genetic algorithm that finds global optimum values for continuous and differentiable functions [60] based on successive generation and transformation of the parameter sets [53]. The DE requires parameters for upper and lower bounds and an objective function for the optimization. During each evolution, in addition to identifying the better parameter sets (or population), the algorithm also introduces a random change to those parameter sets to ultimately get the global optimal parameter values.

The model simulation includes spin up, main (calibration), and post-audit (validation) simulations. Fitness measure statistics, including Nash–Sutcliffe efficiency (NSE), percent bias (PBIAS), root-mean-square error (RMSE), ratio of the RMSE to the standard deviation of measured data (RSR), and Kling–Gupta efficiency (KGE) were used to evaluate the model calibration results. The NSE compares the variance of the residuals (or fitting difference) with the variance of observed lake levels [62]. The PBIAS measures the average residuals or deviations of model results from observed lake levels [63]. The RMSE measures how spread out these residuals from the model results. The RSR is a normalized RMSE by the standard deviation of the observed lake level [64]. The KGE uses the idea of diagnostic decomposition, where the NSE is breakdown into three components (the relative importance of correlation, bias, and variance difference) [63,65]. The KGE ranges from negative infinity to one, with the optimal model prediction having the KGE value close to one.

Where Y^{obs} is observed, Y^{sim} is simulated, Y^{mean} is the mean of the observed lake level change, r is correlation coefficient between the modeled and observed lake levels; γ is a ratio between the standard deviation of modeled and observed lake levels and β is a ratio between the standard deviation and mean of the modeled and observed lake levels (Table 1).

Table 1. Representing statistical model performance criteria.

Statistics	Ranges	Optimal Value
$\text{NSE} = 1 - \left[\frac{\sum_{i=1}^{n}\left(Y_i^{obs} - Y_i^{sim}\right)^2}{\sum_{i=1}^{n}\left(Y_i^{obs} - Y^{mean}\right)^2} \right]$	$-\infty$ to 1	1
$\text{PBIAS} = \left[\frac{\sum_{i=1}^{n}\left(Y_i^{obs} - Y_i^{sim}\right) \times 100}{\sum_{i=1}^{n}\left(Y_i^{obs}\right)} \right]$	0 to 100	0
$\text{RSR} = \frac{\text{RMSE}}{\text{STDEVobs}} = \left[\frac{\sqrt{\sum_{i=1}^{n}\left(Y_i^{Obs} - Y_i^{sim}\right)^2}}{\sqrt{\sum_{i=1}^{n}\left(Y_i^{Obs} - Y^{mean}\right)^2}} \right]$	0 to 1	0
$\text{KGE} = 1 - \sqrt{(r-1)^2 + (\gamma-1)^2 + (\beta-1)^2}$ $\beta = \frac{\mu_s}{\mu_o}$ $\gamma = \frac{CV_s}{CV_o} = \frac{\sigma_s/\mu_s}{\sigma_o/\mu_o}$	0 to 1	1

Manhole overflows were extracted from the calibrated SWMM model and were used as input for the FIRM simulation. The spatial extent of the model is compared to the observed flood regions. Since there was no direct measurement of flood spatial extent and depth, the observed flood area was reconstructed based on pictures of the area taken during the flood event and obtained from social media twitter. The social media information includes both pictures, texts, and street names, allowing us to identify the exact locations and extent of the flood. The previous reports in the cities also indicated that there had been multiple incidences of pluvial flooding near side roads and along intersections. This information was used to delineate the observed flood region, which was then used to validate the flood inundation model. The process of flood boundary delineation is depicted, in Figure 6, which show how the flooded region in the study area was extracted from social media outlets. The images and the text by users are used to identify the exact location of the flooded regions. We used google-earth and high-resolution LiDAR data to determine the relative topographical variation and delineated the inferred flood boundary by considering 360-street and panoramic view. We also used the city of Edmonds GIS-dataset to correlate the inferred flood boundary and building footprints to check if there exist mismatch between the ground and the street boundaries.

2.4. Model Inundation Accuracy

The model's ability to detect the spatial accuracy of the flood extent was evaluated based on the true positive rate (TPR), the positive predictive value (PPV), the modified fit (MF), and the modified bias (MB) methods. The TPR and PPV are derived from the confusion matrix [17], which is a 2-by-2 matrix containing the TPR and PPV for gridded simulated and observed flood conditions (Table 2). These statistics were used recently to assess the flood inundation model performance in [39,66,67]. TPR measures how well the modeled flood region replicates the observed flood boundary. The maximum TPR (100%) represents that the model fully captures the observed flooded regions. The TPR indicates the model tendency to under-predict the flood hazard [39,66,67]. PPV measures how well the model captures flooded in the model but dry in observation. The value ranges from 0%, indicating over prediction of the flood extent, to 100% for accurately captured the observed boundary.

$$\text{TPR} = \frac{\text{TP}}{\text{TP} + \text{FN}} \times 100, \qquad (3)$$

$$\text{PPV} = \frac{\text{TP}}{\text{TP} + \text{FP}} \times 100, \qquad (4)$$

Table 2. A confusion matrix for flood performance.

	Flooded in Observed Boundary	Dry in Observed Boundary
Flooded in FIRM	True flood (TP)	False flood (FP)
Dry in FIRM	False dry (FN)	True dry (TN)

Flood inundation and recession model (FIRM); true positive rate (TPR); the positive predictive value (PPV); true positive (TP); false negative (FN); false positive (FP); true negative (TN).

Where TP represents the flooded regions in both the observation and model simulation, FP represents the flooded region in the model but dry in the observation, and FN represents regions flooded in reality (i.e., in observation) but simulated as dry. The TPR and PPV percentages represent the overlapping rate between simulated and observed flood areas. Higher percentage values of TPR and PPV indicate higher accuracy of the flood inundation model. The two statistics must be used in combination to measure the accuracy of the model since they each evaluate the different performance of the model. Specifically, the TPR and PPV measure how well the model captures the observed flood and flooded pixel that are dry in observation, respectively. For example, the TRP value can be 100% if the model captures all the observed flooded cells even though it may also consider some dry cells as flooded (refer to Equation (3)). Similarly, the PPV value can be 100% if the model captures all the dry cell even though it may also consider some flooded cell as dry.

Other methods of model performance evaluation are fit and bias indicators, which are also commonly used for flood inundation modeling [40,44,68–70]. Previous studies used both fit and bias indicators to compare flood inundation extents between different models. For this research, we infer observed flood extents based on street photos taken during the actual flood event and compare them with the flood inundation results from our model. The modified fit indicator is calculated based on overlapping areas between observed and simulated inundated areas. The indicator ranges from 0% to 100% for poor and ideal model performances, respectively. While the modified bias indicates the overall difference between the simulated and observed flood extents. Positive and negative modified biases indicate an overestimation and underestimation of flood extents by the model, respectively.

$$\text{Modified fit} = \frac{TP}{TP + FP + FN} \times 100, \quad (5)$$

$$\text{Modified bias} = \left(\frac{TP + FP}{TP + FN} - 1\right) \times 100, \quad (6)$$

3. Results and Discussion

3.1. SWMM Model Calibration and Validation

We compared simulated versus observed lake level changes, and show our model captures the lake level fluctuations reasonably well (Figures 7–9). The main objective of the SWMM simulation was to estimate the flood condition in Edmonds from the 19 September, 2016 storm event. For model stability and to represent preexisting hydrological conditions such as soil moisture content, we considered ten months (from 1 August, 2015 to 31 May 2016) of simulation as a spin up. The model calibrated using data from 1 June 2015 to 31 January 2018; and validated using data from 1 February, 2018 to 10 January 2019. After the sensitivity analysis using the manual calibration, we identified five parameters for model calibration. These includes the imperviousness percentage, width, roughness coefficient, depression storage, and the hydraulic conductivity of the soil. The calibration is performed in the DEoptim R-package. Table 3 indicates the upper and lower limits of the model parameters used in the calibration process.

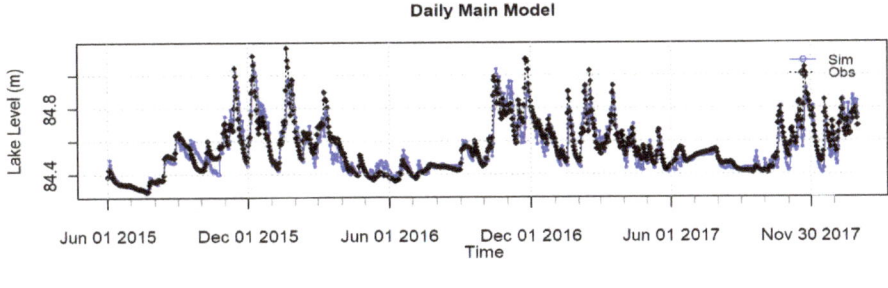
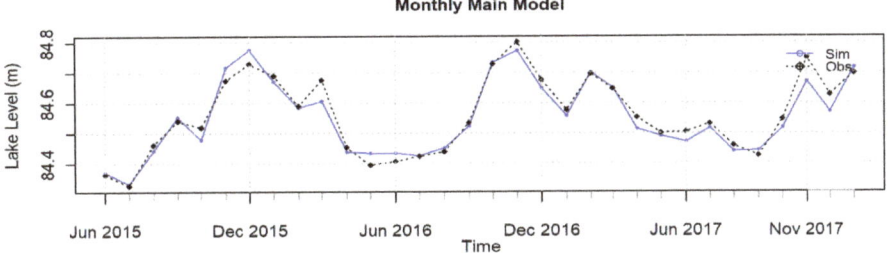

Figure 7. Daily and monthly simulated and observed lake level change for the calibration period.

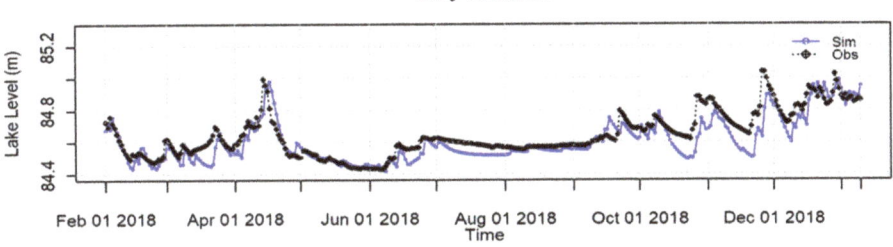
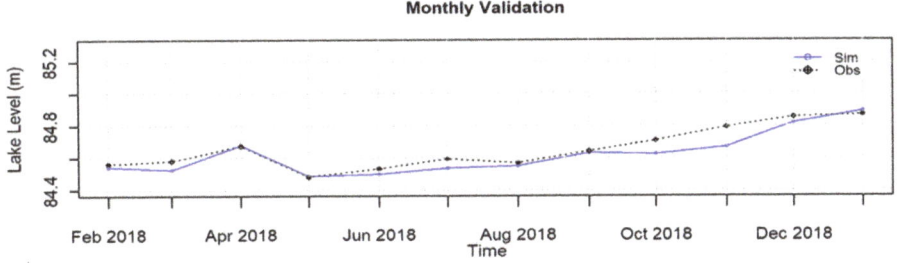

Figure 8. Daily and monthly simulated and observed lake level change for the validation period.

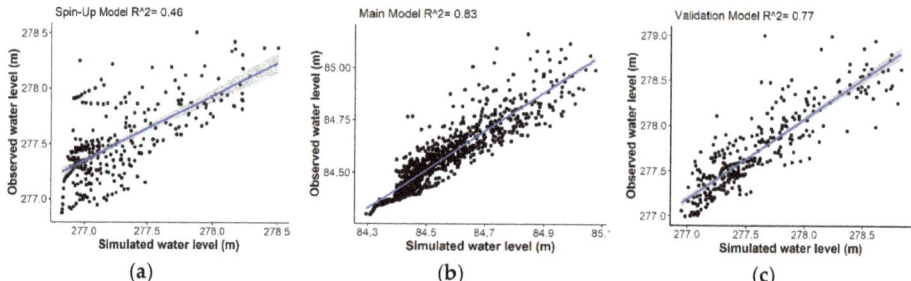

Figure 9. Scatter plots of observed and simulated lakes level during the spin up (**a**), calibration (**b**), and validation (**c**) periods. The lines are the linear regression fit with 95% confidence intervals.

Table 3. Model parameters upper and lower limits, as well as their optimal values.

Parameters	Lower–Upper Bound	Optimal Values
Impervious (%)	25–90	70
Width (m)	150–300	152
Roughness (–)	0.01–0.03	0.012
Depression Storage (mm)	1.2–5.2	1.78
Hydraulic Conductivity (mm/h)	0.1–3	0.11

Figures 7 and 8 shows the comparison of the model simulation and observed time series plots for daily and monthly average lake level change for the calibration and validation simulation periods, respectively. The figures demonstrate that the lake level changes as a result of storm events were reasonably captured for the calibration period. The validation results for both daily and monthly observed and simulated lake levels show the ability of the model to predict beyond the calibration period.

Figure 9 indicates the correlation between the observed and simulated lake level change for spin-up, the main model simulation, and the validation period. The regression coefficient (R^2) is 0.42 for the spin up period, 0.83 for the calibration period, and 0.77 for the validation model simulation period. The correlation coefficients for the calibration and validation period also confirm the model captured the observed lake level reasonably well. The simulation was also evaluated using the NSE, KGE, RSR, and PBIAS.

The statistical summary of the spin up, calibration and validation simulations results are presented in Table 4. For the calibration and validation periods, the KGE are 0.91 and 0.88, respectively, while the NSE is 0.82 and 0.67. These indicate satisfactory model performance. The RSR, which indicates the variation in the residuals, is between 0 and 0.5. This is considered a very good performance [71]. The PBIAS values are also close to 0, which confirm that the model simulates the observed water level with minimal bias. Compared to the daily model performance, the monthly model performance is improved.

Table 4. Model performance statistics to evaluate the Storm Water Management Model (SWMM) daily and monthly lake water level simulations.

Simulation	KGE		NSE		RSR		PBIAS		Performance Rating [71]	
	Daily	Mon	Daily	Mon	Daily	Mon	Daily	Mon	Daily	Mon
Spin Up	0.64	0.61	−0.31	−1.15	1.14	RSR	−0.10	−0.10	Unsat *	Unsat *
Calibration	0.91	0.96	0.82	0.94	0.43	1.39	0.00	0.00	V. good ^	V. good ^
Validation	0.88	0.95	0.67	0.81	0.57	0.24	0.00	0.00	Good	V. good ^

Kling–Gupta efficiency (KGE); Nash–Sutcliffe efficiency (NSE); ratio of the RMSE to the standard deviation of measured data (RSR); percent bias (PBIAS); * Unsatisfactory; ^ Very good.

3.2. Flood Inundation and Recession

The spatial extent and depth of the flood inundation were simulated for a pluvial flood event that happened on 19 September 2016 using FIRM. A detailed conveyance network system with sub-daily meteorological data, such as rainfall data, were used to capture the flood event using SWMM continuous simulation. The FIRM simulate the floods caused by overflow from two manholes, and its results were compared with the inferred flood boundary (Figure 10). Figure 10 represents the areal extent and depth of the flood inundation at two locations. The red dot-line represents the inferred flood boundary, and the black point represents manholes. The color represents flooding depth. The fine-resolution LiDAR data were able to identify detailed urban infrastructures, such as building food prints, and streets. The FIRM was able to identify elevated urban infrastructures and with low-lying streets. Overall, coupling both the 1D SWMM model with the 2D FIRM model was able to delineate the spatial extent and depth of the flood generated from manholes overflow in the study area.

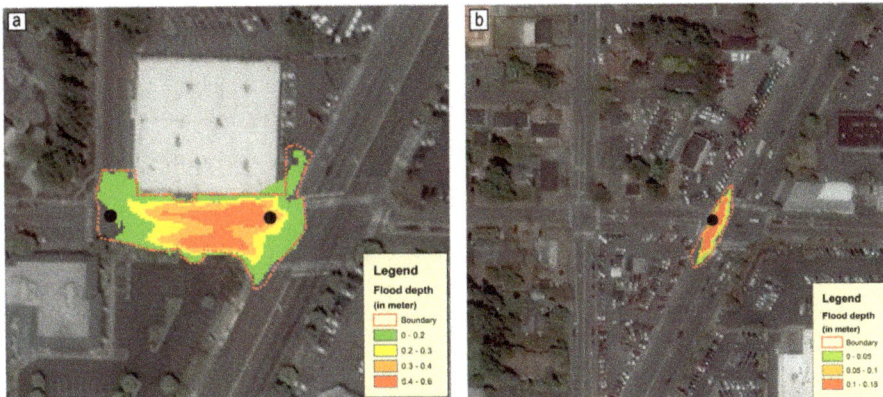

Figure 10. Simulated flood depth and extent (color maps) and observed flood inundation boundaries (red dotted lines). The areal extent and depth of flood in Case 1 (**a**) and Case 2 (**b**).

The flood started from a given manhole and propagated spatially by filling any neighboring grid cells with lower elevations. As shown in Figure 10a, representing Case 1, the flood is concentrated across a highway going west to east. The topographic variations around the manhole are relatively small, with the mean slope of the flooded region being 1.2 degrees (Figure 11a). The flat slope, particularly along the street, enables the overland flow to inundate along the street perpendicular to the main highway. Based on the FIRM result, the areal extent, maximum flood depth and volume of the inundation region is 2129 m^2, 0.6 m, and 710 m^3, respectively. The depth of the flood is controlled by the local elevation and the amount of the excess overland flow. The low-lying part of the street generally has deeper flood depth compared to the peripheral part of the flood extent. Hence, the pavement is often elevated compared with the street elevation. Another manhole flooding in our study (Figure 10b, Case 2) was used to evaluate the flood inundation simulation. In this case, the manhole is located in the steeper part of the street, where the average slope of the street is 5.2 degrees from the west to east (Figure 11b). Consequently, the flood is mostly concentrated along with the highway in the north and south direction, and did not spread much laterally. The volume, areal extent, and maximum depth of the flooded region are 38 m^3, 426 m^2, and 0.15 m, respectively.

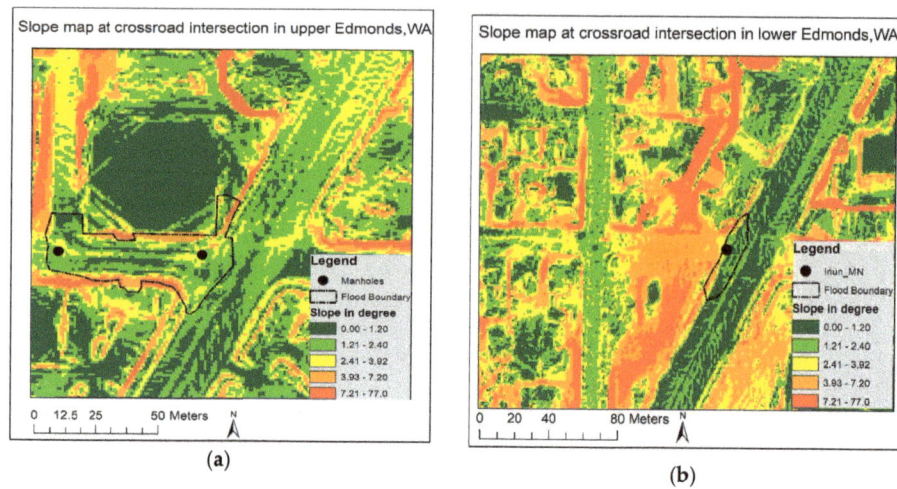

Figure 11. Slope map of the two-flooded regions considered in our study. Case 1 (**a**) and Case 2 (**b**).

Based on the two cases, we observe that the flood inundation model considers the spatial heterogeneity of the surface feature to inundate, for example, flood inundation algorithm was able to differentiate building with streets (Figure 10a) and intersections of streets with varying slope and elevation (Figure 10b). FIRM inundates the lowest level and cover wider area for a relatively gentle region. Conversely, for the manhole located in a steeper region, the algorithm follows the preferred flow direction and inundate relatively smaller area.

To determine the flood recession, we assumed that once the storm ceased and the pipe full capacity decreased, the water eventually drains back to the conveyance system unless it is isolated from the manhole because of depression storages. Accordingly, some of the inundated floodwaters may recess, while the remaining waters are left as ponding associated with local topographic barriers. The result for Case 1 shows that most of the floodwaters drain since the manhole is located at a lower elevation. There is some ponded water away from the manhole due to possible topographic barriers between the manhole and the flooded areas (Figure 12a). Since the manhole is located at a lower elevation, most of the flooded water drained to the manhole. The volume, areal extent, and maximum depth of the ponded region are 15 m^3, 89 m^2, and 0.17 m, respectively. For Case 2, the ponded water is generally concentrated near the manhole due to the local topographical depression around the manhole (Figure 12b). The flood volume, the areal and depth of the ponded water decreases to 3 m^3, 85 m^2, and 0.06 m, respectively. The results showed the FIRM abilities to determine the maximum flood extent and the extent aftermath of a given storm. Each information is important to assess the flooding risk and associated potential short-term (flood inundation) and long-term (flood recession) impacts.

Figure 12. Flood spatial extents aftermath of storm and flood recession into the manholes for Case 1 (**a**) and Case 2 (**b**).

3.3. Model Inundation and Recession Accuracy

In addition to the visual comparison of the observed and simulated flood areas, the model performance evaluation for the inundation was carried out using statistical measures that compare the simulated and observed gridded flood areas. This enables us to identify and assess the model performance based on how accurately the model predicted the observed flooded and dry regions. We adopted the TPR and PPV from [66] and used the modified version of the fitting (MF) and bias (MB) indicators from [44,70]. The model inundation extents are compared with the inferred observed flood boundaries that were extracted from photos of the flood boundaries.

Table 5 shows the model performance indicators used to evaluate the flood inundation model based on the inferred flood regions. The TPR for Case 1 (89%) indicates the model's ability to capture the flooded grid cells, while for Case 2, the TRP is 71%, indicating the model predicted a relatively more flooded region as dry land. Thus, the model under predict the flood hazard in Case 2. The PPV for Case 1 and Case 2 are found to be 95.4% and 97.25%, respectively. These indicate the model's ability to capture the observed none flooded cells. The MF of 85% for Case 1 indicates that the model has better agreements with the observed flood boundary compared to that of Case 2, which has MF of 69.90%, indicating the existence of relatively large variation between the predicted and observed flooded regions. The negative values of the MB for both cases indicates that the flooded regions in both cases were underestimated. Overall, the relative errors are relatively higher for Case 2 compared to Case 1. This is due to underestimating the flood hazard in Case 2 compare with the observed boundary, and possibly due lack of the FIRM to simulate the impact of direct rainfall during the flood events or lack to represent flood water loses into the buildings.

Table 5. Statistical evaluations of the flood inundation model based on inferred flood area at the two manhole locations (Case 1 and Case 2).

Inundation Model Performance	Case 1	Case 2
True positive rate, TPR (%)	89.04	71.31
Positive predictive value, PPV (%)	95.44	97.25
Modified fit, MF (%)	85.04	69.90
Modified bias, MB (%)	−6.71	−26.68

The low TPR values compared to the PPV values suggest that the flood inundation model underestimated the total flooded regions for both cases; but predicted well the flooded region within the observed flood boundaries (Figure 10). The underestimation of the flood areas might be due to the model inundation algorithm not incorporating the additional flooding resulted from direct precipitation or generated surface runoff. The relatively poor performance of the model for Case 2 might also be due to the inundation algorithm limitation to incorporate direct additional flood flux from upstream overland flow into the flooded regions. In addition, the relatively higher slopes in the area, which facilities rapid overland flow from the manhole toward the low elevated regions, may have impacted the model performance. Moreover, the relatively better model performance for Case 1 might be because of the relatively homogenous topography in the area, which is well represented by the 1-m LiDAR data.

Coupling the FIRM model and hydrodynamic model with projected future storm scenarios may help to identify areas that may experience future manhole flooding. This modeling capability can help to better assess flooding risk, and improve designs of storm water drainage systems in flood-prone urban areas. To further improve the work, it is important to consider direct rainfall during storm events and other sources or losses of floodwaters (e.g., possible loses of floodwater by draining into the building or addition of excess runoff from rooftops), as well as the land cover and slopes.

4. Conclusions

We have presented effective flood inundation and recession methodologies that use overflow from given manholes and topography of an urban region. We used the SWWM model to estimate the volume of overflow from manholes. In order to determine the associated flood depth and extent during and after storm events, we developed a flood inundation and recession model (FIRM) that uses high-resolution LIDAR elevation data. SWMM was developed on the basis of watershed characteristics and the drainage conveyance network in the area. SWMM was calibrated using a differential evolution optimization method and validated based on observed lake level data at the outlet of the watershed. The manhole overflows were extracted and used in FIRM to delineate the spatial extent and flood depths.

The spatial extent of the simulated flood area was compared with the observed flood boundary, which was derived from social media pictures and reports from the cities. Two case studies, based on flood events in Edmonds, WA, were considered to evaluate the flood inundation and recession model. In these case studies, the flood occurred across and along a main highway under different topographical characteristics. The results showed that the spatial extent of the flood regions is highly influenced by local topography and the position of the manholes. Particularly, the spatial arrangement of the manhole and the slope of nearby areas are crucial for determining the spatial extent, spatial heterogeneity of the flood depth, and selecting preferential flow paths to inundate low-lying areas. The model is able to capture the flood extent for manhole overland flow in fluvial flood events. Incorporating the direct impact of rainfall on the fluvial flood event can improve the representation of the physical process and the accuracy of the model. As the flood recession observation data are scarce, the performance of the flood recession model result was difficult to quantify. Finally, proper understanding and representation of the study area, the boundary condition, and engineering structures are important for the flood inundation and recession modeling associated with manhole overland flow.

Regional authorities can utilize the presented model (FIRM) by coupling with existing hydrodynamic modeling (e.g., SWMM) to quantify flood hazard based on pluvial generated overland flooding and manholes induced flooding in urban areas, where flood mechanism is complex and modified by local infrastructures. FIRM can be used to estimate the areal extent and depth of flood caused by manholes overland flow during a flood event (flood inundation) and after the event is over (flood recession). Because of the relative simplicity of the model and its uses of readily available data, the model can be used for a real-time assessment of flood progression and to identify potential impact areas. The model's ability to simulate flood recession will also allow identifying areas where the flooded water will remain ponded for days after the floodwater subsides. The ponded water, or the floodwater not drained, can impact human health and properties. In addition to the real-time forecast of flood inundation and estimation of the aftermath ponding condition for the existing drainage system, the model can be used to design better a new or retrofit the current drainage system to minimize the overflow and ponding after the flood events. The model can be used to assess the flood condition under multiple storms, watershed conditions, and drainage scenarios. It can contribute to our understanding of climate change and appropriate engineering designs for mitigation.

Author Contributions: Conceptualization, M.G. and Y.D.; methodology, M.G. and Y.D.; validation, M.G. and Y.D.; writing—original draft preparation, M.G.; writing—review and editing, M.G. and Y.D.; supervision, Y.D.; funding acquisition, Y.D. All authors have read and agreed to the published version of the manuscript.

Funding: Department of Defense's Strategic Environmental Research and Development Program (SERDP) under contract W912HQ-15-C-0023.

Acknowledgments: We are thankful for the city of Edmonds and Mountlake Terrace for providing sewer network dataset and observation data. The first author is grateful for the research input from Joan Wu, Akram Hossain, Jennifer Adam, Mark Wigmosta, Debra Perrone, and Scott Jasechko. We are thankful for the anonymous reviewers for helpful comments on the manuscript.

Conflicts of Interest: The authors declare no conflict of interest. The funders had no role in the design of the study; in the collection, analyses, or interpretation of data; in the writing of the manuscript, or in the decision to publish the results.

References

1. Wallemacq, P.; Herden, C.; House, R. *The Human Cost of Natural Disasters 2015: A Global Perspective*; Technical report; Centre for Research on the Epidemiology of Disasters: Brussels, Belgium, 2015.
2. Stocker, T.F.; Qin, D.; Plattner, G.-K.; Tignor, M.; Allen, S.K.; Boschung, J.; Nauels, A.; Xia, Y.; Bex, V.; Midgley, P.M. Climate change 2013: The physical science basis. In *Contribution of Working Group I to the Fifth Assessment Report of the Intergovernmental Panel on Climate Change*; Cambridge University Press: Cambridge, UK, 2013; Volume 1535.
3. Technical Mapping Advisory Council (TMAC). Technical Mapping Advisory Council (TMAC) 2015 Annual Report Summary. Available online: https://www.fema.gov/media-library-data/1454954186441-34ff688ee1abc00873df80c4d323a4df/TMAC_2015_Annual_Report_Summary.pdf (accessed on 13 March 2020).
4. Federal Emergency Management Agency (FEMA). Loss Dollars Paid by Calendar Year. Available online: https://www.fema.gov/loss-dollars-paid-calendar-year (accessed on 13 March 2020).
5. European Environmental Agency (EEA). Mapping the impacts of natural hazards and technological accidents in Europe – an overview of the last decade. EEA Technical Report No13/2010. Available online: https://www.eea.europa.eu/publications/mapping-the-impacts-of-natural (accessed on 13 March 2020).
6. Wang, W.; Li, H.Y.; Leung, L.R.; Yigzaw, W.; Zhao, J.; Lu, H.; Deng, Z.; Demisie, Y.; Blöschl, G. Nonlinear filtering effects of reservoirs on flood frequency curves at the regional scale. *Water Resour. Res.* **2017**, *53*, 8277–8292. [CrossRef]
7. Ye, S.; Li, H.-Y.; Leung, L.R.; Guo, J.; Ran, Q.; Demissie, Y.; Sivapalan, M. Understanding flood seasonality and its temporal shifts within the contiguous United States. *J. Hydrometeorol.* **2017**, *18*, 1997–2009. [CrossRef]
8. Milner, A.M.; Picken, J.L.; Klaar, M.J.; Robertson, A.L.; Clitherow, L.R.; Eagle, L.; Brown, L.E. River ecosystem resilience to extreme flood events. *Ecol. Evol.* **2018**, *8*, 8354–8363. [CrossRef] [PubMed]
9. Scanlon, B.R.; Keese, K.E.; Flint, A.L.; Flint, L.E.; Gaye, C.B.; Edmunds, W.M.; Simmers, I. Global synthesis of groundwater recharge in semiarid and arid regions. *Hydrol. Processes* **2006**, *20*, 3335–3370. [CrossRef]
10. Wang, X.; Zhang, G.; Xu, Y.J. Impacts of the 2013 extreme flood in Northeast China on regional groundwater depth and quality. *Water* **2015**, *7*, 4575–4592. [CrossRef]
11. Jasechko, S.; Birks, S.J.; Gleeson, T.; Wada, Y.; Fawcett, P.J.; Sharp, Z.D.; McDonnell, J.J.; Welker, J.M. The pronounced seasonality of global groundwater recharge. *Water Resour. Res.* **2014**, *50*, 8845–8867. [CrossRef]
12. Cuthbert, M.O.; Taylor, R.G.; Favreau, G.; Todd, M.C.; Shamsudduha, M.; Villholth, K.G.; MacDonald, A.M.; Scanlon, B.R.; Kotchoni, D.V.; Vouillamoz, J.-M. Observed controls on resilience of groundwater to climate variability in sub-Saharan Africa. *Nature* **2019**, *572*, 230–234. [CrossRef]

13. Dahlke, H.; Brown, A.; Orloff, S.; Putnam, D.; O'Geen, T. Managed winter flooding of alfalfa recharges groundwater with minimal crop damage. *Calif. Agr.* **2018**, *72*, 65–75. [CrossRef]
14. GebreEgziabher, M. An Integrated Hydrogeological Study to Understand the Groundwater Flow Dynamics in Raya Valley Basin, Northern Ethiopia: Hydrochemistry, Isotope Hydrology and Flow Modeling Approaches. Master's Thesis, Addis Ababa University, Addis Ababa, Ethiopia, 2011.
15. Changnon, S.A., Jr. Recent studies of urban effects on precipitation in the United States. *Bull. Am. Meteorol. Soc.* **1969**, *50*, 411–421. [CrossRef]
16. Arnbjerg-Nielsen, K.; Willems, P.; Olsson, J.; Beecham, S.; Pathirana, A.; Bülow Gregersen, I.; Madsen, H.; Nguyen, V.-T.-V. Impacts of climate change on rainfall extremes and urban drainage systems: A review. *Water Sci. Technol.* **2013**, *68*, 16–28. [CrossRef]
17. Boyd, E.; Juhola, S. Adaptive climate change governance for urban resilience. *Urban Stud.* **2015**, *52*, 1234–1264. [CrossRef]
18. Ford, A.; Barr, S.; Dawson, R.; Virgo, J.; Batty, M.; Hall, J. A multi-scale urban integrated assessment framework for climate change studies: A flooding application. *Comput. Environ. Urban* **2019**, *75*, 229–243. [CrossRef]
19. Chen, J.; Hill, A.A.; Urbano, L.D. A GIS-based model for urban flood inundation. *J. Hydrol.* **2009**, *373*, 184–192. [CrossRef]
20. Wang, R.-Q.; Mao, H.; Wang, Y.; Rae, C.; Shaw, W. Hyper-resolution monitoring of urban flooding with social media and crowdsourcing data. *Comput. Geosci.* **2018**, *111*, 139–147. [CrossRef]
21. National Academies of Sciences, Engineering, and Medicine. *Framing the Challenge of Urban Flooding in the United States*; National Academies Press: Washington, DC, USA, 2019.
22. Jacobson, C.R. Identification and quantification of the hydrological impacts of imperviousness in urban catchments: A review. *J. Environ. Manage.* **2011**, *92*, 1438–1448. [CrossRef] [PubMed]
23. Zhang, W.; Villarini, G.; Vecchi, G.A.; Smith, J.A. Urbanization exacerbated the rainfall and flooding caused by hurricane Harvey in Houston. *Nature* **2018**, *563*, 384–388. [CrossRef] [PubMed]
24. Shuster, W.D.; Bonta, J.; Thurston, H.; Warnemuende, E.; Smith, D. Impacts of impervious surface on watershed hydrology: A review. *Urban Water J.* **2005**, *2*, 263–275. [CrossRef]
25. Diakakis, M.; Deligiannakis, G.; Pallikarakis, A.; Skordoulis, M. Identifying elements that affect the probability of buildings to suffer flooding in urban areas using Google Street View. A case study from Athens metropolitan area in Greece. *Int. J. Disaster Risk Reduct.* **2017**, *22*, 1–9. [CrossRef]
26. Golz, S.; Schinke, R.; Naumann, T. Assessing the effects of flood resilience technologies on building scale. *Urban Water J.* **2015**, *12*, 3043. [CrossRef]
27. Hu, M.; Sayama, T.; Zhang, X.; Tanaka, K.; Takara, K.; Yang, H. Evaluation of low impact development approach for mitigating flood inundation at a watershed scale in China. *J. Environ. Manage.* **2017**, *193*, 430–438. [CrossRef]
28. Brody, S.; Sebastian, A.; Blessing, R.; Bedient, P. Case study results from southeast Houston, Texas: Identifying the impacts of residential location on flood risk and loss. *J. Flood Risk Manage.* **2018**, *11*, S110–S120. [CrossRef]
29. Teng, J.; Jakeman, A.J.; Vaze, J.; Croke, B.F.; Dutta, D.; Kim, S. Flood inundation modelling: A review of methods, recent advances and uncertainty analysis. *Environ. Modell. Softw.* **2017**, *90*, 201–216. [CrossRef]
30. Yu, D.; Coulthard, T.J. Evaluating the importance of catchment hydrological parameters for urban surface water flood modelling using a simple hydro-inundation model. *J. Hydrol.* **2015**, *524*, 385–400. [CrossRef]
31. Courty, L.G.; Pedrozo-Acuña, A.; Bates, P.D. Itzï (version 17.1): An open-source, distributed GIS model for dynamic flood simulation. *Geosci. Model. Dev.* **2017**, *10*, 1835. [CrossRef]
32. Rosenberg, E.A.; Keys, P.W.; Booth, D.B.; Hartley, D.; Burkey, J.; Steinemann, A.C.; Lettenmaier, D.P. Precipitation extremes and the impacts of climate change on stormwater infrastructure in Washington State. *Clim. Change* **2010**, *102*, 319–349. [CrossRef]
33. Mishra, V.; Ganguly, A.R.; Nijssen, B.; Lettenmaier, D.P. Changes in observed climate extremes in global urban areas. *Environ. Res. Lett.* **2015**, *10*, 024005. [CrossRef]
34. Muis, S.; Güneralp, B.; Jongman, B.; Aerts, J.C.; Ward, P.J. Flood risk and adaptation strategies under climate change and urban expansion: A probabilistic analysis using global data. *Sci. Total Environ.* **2015**, *538*, 445–457. [CrossRef]
35. Zhao, G.; Xu, Z.; Pang, B.; Tu, T.; Xu, L.; Du, L. An enhanced inundation method for urban flood hazard mapping at the large catchment scale. *J. Hydrol.* **2019**, *571*, 873–882. [CrossRef]

36. Zhao, T.; Shao, Q.; Zhang, Y. Deriving flood-mediated connectivity between river channels and floodplains: Data-driven approaches. *Sci. Rep.* **2017**, *7*, 43239. [CrossRef]
37. Wang, X.; Kinsland, G.; Poudel, D.; Fenech, A. Urban flood prediction under heavy precipitation. *J. Hydrol.* **2019**, *577*, 123984. [CrossRef]
38. Jamali, B.; Bach, P.M.; Cunningham, L.; Deletic, A. A Cellular Automata Fast Flood Evaluation (CA-ffé) Model. *Water Resour. Res.* **2019**, *55*, 4936–4953. [CrossRef]
39. Zheng, X.; Maidment, D.R.; Tarboton, D.G.; Liu, Y.Y.; Passalacqua, P. GeoFlood: Large-Scale Flood Inundation Mapping Based on High-Resolution Terrain Analysis. *Water Resour. Res.* **2018**, *54*, 10013–10033. [CrossRef]
40. Yang, T.-H.; Chen, Y.-C.; Chang, Y.-C.; Yang, S.-C.; Ho, J.-Y. Comparison of different grid cell ordering approaches in a simplified inundation model. *Water* **2015**, *7*, 438–454. [CrossRef]
41. Meng, X.; Zhang, M.; Wen, J.; Du, S.; Xu, H.; Wang, L.; Yang, Y. A Simple GIS-Based Model for Urban Rainstorm Inundation Simulation. *Sustainability* **2019**, *11*, 2830. [CrossRef]
42. Sörensen, J.; Mobini, S. Pluvial, urban flood mechanisms and characteristics–assessment based on insurance claims. *J. Hydrol.* **2017**, *555*, 51–67. [CrossRef]
43. Leandro, J.; Martins, R. A methodology for linking 2D overland flow models with the sewer network model SWMM 5.1 based on dynamic link libraries. *Water Sci. Technol.* **2016**, *73*, 3017–3026. [CrossRef]
44. Son, A.-L.; Kim, B.; Han, K.-Y. A simple and robust method for simultaneous consideration of overland and underground space in urban flood modeling. *Water* **2016**, *8*, 494. [CrossRef]
45. Chang, T.-J.; Wang, C.-H.; Chen, A.S.; Djordjević, S. The effect of inclusion of inlets in dual drainage modelling. *J. Hydrol.* **2018**, *559*, 541–555. [CrossRef]
46. Jang, J.-H.; Chang, T.-H.; Chen, W.-B. Effect of inlet modelling on surface drainage in coupled urban flood simulation. *J. Hydrol.* **2018**, *562*, 168–180. [CrossRef]
47. Seyoum, S.D.; Vojinovic, Z.; Price, R.K.; Weesakul, S. Coupled 1D and noninertia 2D flood inundation model for simulation of urban flooding. *J. Hydraul. Eng.* **2012**, *138*, 23–34. [CrossRef]
48. Chen, A.S.; Leandro, J.; Djordjević, S. Modelling sewer discharge via displacement of manhole covers during flood events using 1D/2D SIPSON/P-DWave dual drainage simulations. *Urban Water J.* **2016**, *13*, 830–840. [CrossRef]
49. Rossman, L.A.; Huber, W. Storm water management model reference manual volume II–hydraulics. *US Environ. Prot. Agency II (Mayo)* **2017**, *190*. Available online: https://nepis.epa.gov/Exe/ZyPDF.cgi?Dockey=P100S9AS.pdf (accessed on 17 April 2020).
50. Kessler, R. Stormwater strategies: Cities prepare aging infrastructure for climate change. *Environ. Health Perspect.* **2011**, *119*, 514–519. [CrossRef] [PubMed]
51. Milly, P.C.; Betancourt, J.; Falkenmark, M.; Hirsch, R.M.; Kundzewicz, Z.W.; Lettenmaier, D.P.; Stouffer, R.J. Stationarity is dead: Whither water management? *Science* **2008**, *319*, 573–574. [CrossRef] [PubMed]
52. Yan, H.; Sun, N.; Wigmosta, M.; Skaggs, R.; Hou, Z.; Leung, L.R. Next-generation intensity–duration–frequency curves to reduce errors in peak flood design. *J. Hydrol. Eng.* **2019**, *24*, 04019020. [CrossRef]
53. Mullen, K.; Ardia, D.; Gil, D.L.; Windover, D.; Cline, J. DEoptim: An R package for global optimization by differential evolution. *J. Stat. Softw.* **2011**, *40*, 1–26. [CrossRef]
54. Means, J.E.; Acker, S.A.; Harding, D.J.; Blair, J.B.; Lefsky, M.A.; Cohen, W.B.; Harmon, M.E.; McKee, W.A. Use of large-footprint scanning airborne lidar to estimate forest stand characteristics in the Western Cascades of Oregon. *Remote Sens. Environ.* **1999**, *67*, 298–308. [CrossRef]
55. Leutnant, D.; Döring, A.; Uhl, M. swmmr-an R package to interface SWMM. *Urban Water J.* **2019**, *16*, 68–76. [CrossRef]
56. Niazi, M.; Nietch, C.; Maghrebi, M.; Jackson, N.; Bennett, B.R.; Tryby, M.; Massoudieh, A. Storm water management model: Performance review and gap analysis. *J. Sustain. Water Built Env.* **2017**, *3*, 04017002. [CrossRef]
57. Gray, J.E.; Pribil, M.J.; Van Metre, P.C.; Borrok, D.M.; Thapalia, A. Identification of contamination in a lake sediment core using Hg and Pb isotopic compositions, Lake Ballinger, Washington, WA, USA. *J. Appl. Geochem.* **2013**, *29*, 1–12. [CrossRef]
58. Thapalia, A.; Borrok, D.M.; Van Metre, P.C.; Musgrove, M.; Landa, E.R. Zn and Cu isotopes as tracers of anthropogenic contamination in a sediment core from an urban lake. *Environ. Sci. Technol.* **2010**, *44*, 1544–1550. [CrossRef]

59. Boyle, D.P.; Gupta, H.V.; Sorooshian, S. Toward improved calibration of hydrologic models: Combining the strengths of manual and automatic methods. *Water Resour. Res.* **2000**, *36*, 3663–3674. [CrossRef]
60. Price, K.; Storn, R.M.; Lampinen, J.A. *Differential Evolution: A Practical Approach to Global Optimization*; Springer Science & Business Media: Berlin, Germany, 2006.
61. Ardia, D.; Mullen, K.; Peterson, B.; Ulrich, J. DEoptim': Differential Evolution in 'R'. Version 2.2-3. 2015. Available online: https://cran.r-project.org/web/packages/DEoptim/DEoptim.pdf (accessed on 4 April 2020).
62. Nash, J.E.; Sutcliffe, J.V. River flow forecasting through conceptual models part I—A discussion of principles. *J. Hydrol.* **1970**, *10*(3), 282–290. [CrossRef]
63. Gupta, H.V.; Kling, H.; Yilmaz, K.K.; Martinez, G.F. Decomposition of the mean squared error and NSE performance criteria: Implications for improving hydrological modelling. *J. Hydrol.* **2009**, *377*, 80–91. [CrossRef]
64. Legates, D.R.; McCabe, G.J., Jr. Evaluating the use of "goodness-of-fit" measures in hydrologic and hydroclimatic model validation. *Water Resour. Res.* **1999**, *35*, 233–241. [CrossRef]
65. Kling, H.; Fuchs, M.; Paulin, M. Runoff conditions in the upper Danube basin under an ensemble of climate change scenarios. *J. Hydrol.* **2012**, *424*, 264–277. [CrossRef]
66. Wang, Y.; Chen, A.S.; Fu, G.; Djordjević, S.; Zhang, C.; Savić, D.A. An integrated framework for high-resolution urban flood modelling considering multiple information sources and urban features. *Environ. Modell Softw.* **2018**, *107*, 85–95. [CrossRef]
67. Wing, O.E.; Bates, P.D.; Sampson, C.C.; Smith, A.M.; Johnson, K.A.; Erickson, T.A. Validation of a 30 m resolution flood hazard model of the conterminous United States. *Water Resour. Res.* **2017**, *53*, 7968–7986. [CrossRef]
68. Bates, P.D.; De Roo, A. A simple raster-based model for flood inundation simulation. *J. Hydrol.* **2000**, *236*, 54–77. [CrossRef]
69. Bernini, A.; Franchini, M. A rapid model for delimiting flooded areas. *Water Resour Manag.* **2013**, *27*, 3825–3846. [CrossRef]
70. Lhomme, J.; Sayers, P.; Gouldby, B.; Samuels, P.; Wills, M.; Mulet-Marti, J. Recent development and application of a rapid flood spreading method. In Proceedings of the FloodRisk 2008 Conference, Oxford, UK, 30 September–2 October 2008; Taylor and Francis Group: London, UK, 2008.
71. Moriasi, D.N.; Arnold, J.G.; Van Liew, M.W.; Bingner, R.L.; Harmel, R.D.; Veith, T.L. Model evaluation guidelines for systematic quantification of accuracy in watershed simulations. *Trans. ASABE* **2007**, *50*, 885–900. [CrossRef]

 © 2020 by the authors. Licensee MDPI, Basel, Switzerland. This article is an open access article distributed under the terms and conditions of the Creative Commons Attribution (CC BY) license (http://creativecommons.org/licenses/by/4.0/).

Article

CFD Modelling of the Transport of Soluble Pollutants from Sewer Networks to Surface Flows during Urban Flood Events

Md Nazmul Azim Beg [1,2,*], Matteo Rubinato [3], Rita F. Carvalho [2] and James D. Shucksmith [4]

1. Tulane River and Coastal Center, Tulane University, 1370 Port of New Orleans Pl, New Orleans, LA 70130, USA
2. MARE—Marine and Environmental Research Centre, Department of Civil Engineering, University of Coimbra, R. Luís Reis Santos, Polo 2, 3030-788 Coimbra, Portugal; ritalmfc@dec.uc.pt
3. School of Energy, Construction and Environment & Centre for Agroecology, Water and Resilience, Coventry University, Coventry CV1 5FB, UK; matteo.rubinato@coventry.ac.uk
4. Department of Civil and Structural Engineering, the University of Sheffield, Mappin Street, Sheffield S1 3JD, UK; j.shucksmith@sheffield.ac.uk
* Correspondence: mbeg@tulane.edu; Tel.: +1-504-405-1741

Received: 3 August 2020; Accepted: 1 September 2020; Published: 9 September 2020

Abstract: Surcharging urban drainage systems are a potential source of pathogenic contamination of floodwater. While a number of previous studies have investigated net sewer to surface hydraulic flow rates through manholes and gullies during flood events, an understanding of how pollutants move from sewer networks to surface flood water is currently lacking. This paper presents a 3D CFD model to quantify flow and solute mass exchange through hydraulic structures featuring complex interacting pipe and surface flows commonly associated with urban flood events. The model is compared against experimental datasets from a large-scale physical model designed to study pipe/surface interactions during flood simulations. Results show that the CFD model accurately describes pipe to surface flow partition and solute transport processes through the manhole in the experimental setup. After validation, the model is used to elucidate key timescales which describe mass flow rates entering surface flows from pipe networks. Numerical experiments show that following arrival of a well-mixed solute at the exchange structure, solute mass exchange to the surface grows asymptotically to a value equivalent to the ratio of flow partition, with associated timescales a function of the flow conditions and diffusive transport inside the manhole.

Keywords: pollutant transport; hydraulic structures; urban flooding; urban drainage; CFD

1. Introduction

Urban flooding events can cause significant economic and societal disruption. Numerous studies [1–3] have suggested that the occurrence of flooding in urban areas is likely to increase in the future due to increased urbanisation and changes in precipitation patterns, making intense rainfall events and the inundation of local drainage systems more common. The majority of urban flooding hazard studies focus on the economic damage, or direct risks to the public derived from hydraulic modelling of the depth and velocity of floodwaters resulting from historic or design rainfall events (see, e.g., in [4,5]). However, an increasing number of studies have also considered the public health risks of exposure to flood water, which may take the form of long term mental impacts [6], or illness from direct exposure of the public to contaminated flood water. Urban floodwater may contain a mix of rainwater, stormwater runoff and waste/foul water from surcharging urban drainage systems and therefore may contain harmful bacteria [7,8]. For example, ten Veldhuis et al. [9] sampled and analysed flood water from three urban flooding incidents in the Hague, the Netherlands in areas served by combined sewers. In the study, values of intestinal enterococci and *E. coli* were found to be 1 to 3 orders of magnitude higher than values for good bathing water quality according to the EU Directive 2006/7/EC.

Understanding the concentrations, transport and fate of harmful contaminants in urban floodwaters for effective health risk assessment is challenging [10]. Current state of the art urban flood risk models consider urban hydrological processes and utilise hydrodynamic principles to route resulting flows in both piped drainage and surface overland systems, with interaction (i.e., mass transfer) nodes such as manholes or gullies, which are commonly represented by weir or orifice equations [11–13]. Although flood model calibration and validation is often difficult due to a paucity of full scale data, such tools are generally considered to give tolerable predictions of flood depths and are widely used for risk evaluation and asset management [14]. Recently, Mark et al. [15] developed an approach to integrate an understanding of contaminant transport and health risk into flood models, utilising the 2D Advection Dispersion Equation to simulate the mixing and transport of wastewater surcharging from drainage systems within overland surface flow (assuming a constant pathogen level within the surcharging flow). However, such approaches can significantly simplify a number of processes concerning sources, transport, survival and transformations of harmful contaminants (e.g., see in [10]). The number of additional terms and associated parameters required to account for transport and fate processes exacerbate non-identifiability and equifinality issues which are a common problem for complex integrated models [16]. To develop a more robust understanding of health risks posed by urban flood waters, detailed information is required concerning individual processes associated with sources, transport pathways and life cycles of pathogens from sewer/drainage networks to surface flows and on urban surfaces. For example, recent studies have considered the behaviour of waterborne pathogens on different urban surfaces [17] and evaluation of pathogen levels in urban rainfall runoff flows [18].

However, as far as the authors are aware, no studies to date have considered the exchange of contaminated material (in soluble or particulate form) from drainage/sewer networks to surface flows during flood events via interaction structures such as gullies and/or manholes. Flows in and around surcharging hydraulic structures are highly complex and three-dimensional, especially during interactions with surface flood flows [19]. It is also likely that contamination concentrations within urban drainage/sewer networks will vary significantly as the proportion of stormwater and quantity and nature of contaminated material (i.e., dissolved vs. entrained solids) within the network varies during flood events. Numerous studies have considered the mixing of soluble material in manhole structures in the absence of interacting surface flood flows, demonstrating that mixing/transport (and thus mass exchange) processes are sensitive to geometrical characteristics and poorly described using commonly used simple models such as the 1D ADE which are commonly used to model pollutant transport and mixing in piped networks (see, e.g., in [20,21]). More complex 3D CFD based approaches have been shown to be able to quantify hydraulic and solute mixing processes in hydraulic structures such as manholes [12,22–25]. However, to date such models have not been experimentally validated in urban flood situations which include complex interactions between piped and surface flows [19]. While such 3D models are too computationally expensive to be used in direct design or network simulation, validated CFD models can be used to conduct experiments which may elucidate relationships and timescales describing the transport of materials to surface flows, understand the influence of geometric or hydraulic variables on mixing and mass transport characteristics or be used to calibrate simpler models.

Understanding how contaminants move from sewer networks to surface flows is a key aspect for understanding health risks posed by urban floods and possibly to foster the design of techniques to mitigate negative effects. This study conducts a detailed 3D numerical simulation of flow and soluble mass transport through a manhole during surface flooding conditions where net sewer to surface exchange flows are simulated. Whilst the focus of this study is limited to soluble pollutants only (i.e., those fully dissolved in the flow), it is recognised that the transport of contaminated solid material (e.g., fine sewer sediments) is also relevant in this context. The aims of the paper are to (1) compare the model outputs to new hydraulic and solute transport experimental datasets collected in a scale model facility designed to study interactions between pipe and surface flows. (2) Conduct numerical experiments to provide a more complete understanding of mass exchange to surface flows via hydraulic structures, including characteristic timescales associated with the occurrence of steady mass flow rate conditions.

2. Materials and Methods

Section 2.1 presents details of the setup used to gather experimental data to evaluate the numerical model. Section 2.2 provides a definition of key timescales and processes to be explored using CFD modelling and Section 2.3 describes numerical model and tests undertaken.

2.1. Experimental Setup

To collect data required for evaluating the numerical model, an experimental testing campaign was conducted using a physical 1:6 scaled model of a linked sewer/surface system, constructed at the University of Sheffield (Figure 1) [11,19,26–31]. The model is composed of a surface "floodplain" 8.2 m long, 4 m wide, constructed from acrylic (slope of 0.001 m/m). This floodplain is connected to a piped sewer system via a manhole with a diameter of 0.240 m (simulating a 1.440 m manhole at full scale, a size typical of UK urban drainage systems for pipes diameters up to 900 mm [32]). The sewer comprises a 0.075 m (internal) diameter clear acrylic pipe (simulating a 0.450 m pipe at full scale). To simulate flooding conditions, a series of steady flows were passed into the inlets at the upstream boundary of the sewer system and the floodplain. During each test, a portion of the flow within the piped network passed into the surface system via the manhole structure, with the remaining flow passing to the pipe outlet tank via the downstream boundary. The scheme of the facility is displayed in Figure 1.

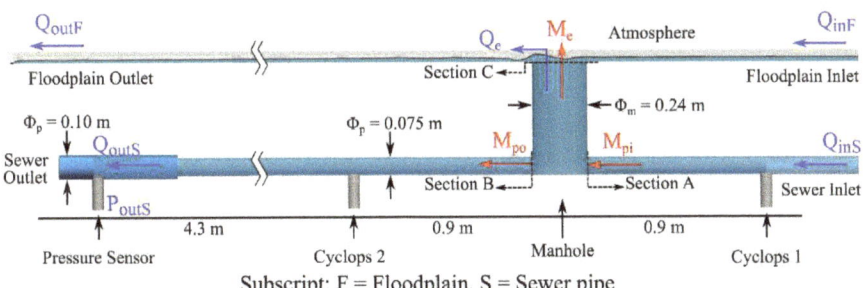

Figure 1. Scheme of the exchange structure showing the floodplain, sewer pipe and the manhole.

The experimental facility was equipped with three electromagnetic flowmeters (two of them at the sewer and surface flow inlet—Q_{InS} and Q_{InF}—and one in the outlet of the sewer—Q_{OutS}) of 0.075 m internal diameter. The accuracy of the flow meters was validated using volumetric discharge readings at the laboratory measurement tank. A butterfly flow control valve was fitted to the pipe that feeds the sewer and the floodplain, calibrated such that steady inflows from 1 to 11 L/s can be set. Electromagnetic flowmeters and butterfly valves were monitored and controlled via Labview™ software. For all the tests conducted, flows were first established and allowed to stabilise before data values were recorded. Once established, data were collected for a period of 3 min to define reliable temporally averaged values for each flow mater. Mean steady state flow exchange rate through the manhole structure (Q_e) was quantified based on mass conservation principles (i.e., $Q_e = Q_{InS} - Q_{OutS}$). During the experimental campaign, water column pressure at the sewer outlet point was measured using pressure sensor (Figure 1). This sensor was calibrated to directly convert the output signal (mA) to gauge pressure and this procedure was conducted using a pointer gauge. The measure values were compared against defined calibration outcomes and errors were quantified to be ±0.69 mm within the water depth range of 0 to 600 mm. Values recorded with the pressure sensor were then gathered in real time by using the same Labview™ software described previously.

Experiments to understand solute transport and mass exchange were undertaken by injection of a neutrally buoyant soluble fluorescing dye (Rhodamine WT) into the sewer pipe >8 m upstream of the first measurement point (Cyclops 1 in Figure 1). The distance between the location of the injection and the measurement areas was higher than 10D (D = sewer pipe diameter) to allow cross sectional mixing [33]. Measurement of concentration vs. time profiles upstream and downstream of the manhole was conducted using Cyclops-7F™ fluorimeters. For this experiment, dye of concentration 10^{-3} mg/L was fed into a constant head tank, from where injection into the sewer pipe was controlled by a manual open/close valve. For each test conducted, a 15 s duration pulse of dye of was introduced into the inflow pipe, and the resulting in-pipe concentrations monitored using the fluorimeters. The electrical sensor output was converted to concentration using experimentally predetermined calibration equations.

Experimental tests were conducted under steady state hydraulic conditions over a range of sewer inflows (Q_{InS}) and surface inflows (Q_{InF}), producing different flow exchange rates (Q_e). Reynolds number (Re) for these tests ranged from 1.37×10^6 to 1.72×10^6 in the sewer inlet, which indicates a fully turbulent flow condition. Surface flow depths measured 350 mm upstream of the centreline of the manhole ranged between 5 to 17 mm over the tests conducted. Full details of these test conditions along with their numerical replication in CFD are presented in Table 1 in Section 3.

2.2. Timescales and Mass Exchange Processes

For a given pulse of soluble contaminant passing within a pipe network entering an exchange structure (e.g., a manhole) during sewer-to-surface flow exchange conditions (i.e., $Q_e > 0$), a proportion will pass through the structure remaining within the pipe network and a proportion will exit to the surface flow. The change in total solute mass within the exchange structure at a given point in time can be expressed as

$$\frac{dM_m}{dt} = \dot{M}_{PI} - \dot{M}_e - \dot{M}_{PO} \qquad (1)$$

where $\frac{dM_m}{dt}$ is the rate of change in mass of solute within the exchange structure (mg/s), $\dot{M}_{PI}$ is the solute mass flow rate entering the exchange structure via the pipe network (mg/s), $\dot{M}_e$ is solute mass flow rate (mg/s) leaving the exchange structure to the surface flow and $\dot{M}_{PO}$ is the solute mass flow rate leaving the exchange structure via the pipe network (mg/s).

Considering the arrival of a well-mixed solute of concentration (C_{PI}) at the inlet to the exchange structure under steady inflow conditions and Equation (1) above, a number of characteristic timescales can be defined.

- From time t_o to t_1, solute mass is entirely stored within the exchange structure (prior to solute reaching an exit), thus $\dot{M}_e = \dot{M}_{PO} = 0$.
- Assuming typical dimensions and flow conditions encountered within urban drainage exchange structures such as manholes, between t_1 and t_2, solute mass initially leaves the exchange structure via the outlet pipe only, hence $\dot{M}_e = 0$.
- Between t_2 and t_3 solute mass leaves the exchange structure via the pipe outlet and to the surface, solute mass flow rate to the surface will be dependent on the hydraulic characteristics and evolution of solute inside the exchange structure and all terms in (1) should be considered.
- After t_3 concentration gradients within the structure will have significantly reduced and hence steady mass flow conditions in the structure are achieved, $\frac{dM_m}{dt} = 0$. Considering that the solute

mass flow through an inlet/outlet is a product of the rate of hydraulic flow rate and mean solute concentration, the proportion of solute mass exchanging though each outlet will become equivalent to the flow partition through the structure (Equations (2) and (3)).

$$\frac{\dot{M}_e}{\dot{M}_{PI}} = \frac{Q_e}{Q_{PI}} \qquad (2)$$

$$\frac{\dot{M}_{PO}}{\dot{M}_{PI}} = \frac{Q_{PO}}{Q_{PI}} \qquad (3)$$

Further experiments on specific hydraulic structures are required to understand the characteristic timescales ($t_{1,2,3}$) and how these are affected by local flow characteristics. Flow structures and mixing processes in tanks and urban drainage structures have been studied previously but in the absence of surface flow interaction. For example, a general description of flow structures in manholes under surcharged pipe conditions is given in [12]. When the sewer pipe inflow enters the manhole, three distinguished flow zone can be commonly observed [12,23,34]. A part of the pressurised flow, known as the diffusion zone, expands inside the manhole at a ratio of 1:5 towards the manhole diameter length. The remaining strong velocity zone forms a conical shape which has the same central axis as the inlet pipe. The slope of this cone is generally 1:6.2 towards the manhole length and travels through the manhole diameter towards the outlet. This conical form may create different distinctive scenarios based on the manhole to sewer pipe diameter ratio (ϕ_m/ϕ_m) and available surcharge depth (s). For $3.0 < \phi_m/\phi_p < 4.5$ and with $s > 0.2\phi_m$, the core velocity region travels out of the manhole without contributing to the mixing process [23,35]. This is the most conventional size and surcharge depth characteristics for an overflowing manhole commonly seen the drainage systems and corresponds to the present study. In these cases, the diffusion zone is mainly responsible for solute mixing inside a manhole [20,23]. Part of the diffusion zone interacts with the manhole wall and travels upward. Later, this upward moving flow further divides in two components of which the first part exits through the surface and the last part recirculates within the manhole. However, how these structures interact with a surface flow, how effective they are in transporting solute mass to the surface and key timescales for well-mixed conditions (i.e., Equations (2) and (3)) are currently unclear and will be analysed in the current work using CFD techniques.

2.3. CFD Modelling

The hydraulics of the experimental model was reproduced using three-dimensional CFD modelling tools OpenFOAM® v.18.12 within interFoam solver [36–38], which considers the two-fluid system as isothermal, incompressible and immiscible utilising a Volume of Fluid (VOF) model [39]. Despite Larger Eddy Simulation (LES) models being known to model the turbulence structures of the flow more effectively, LES models are significantly more computationally expensive than those of RANS models. Moreover, RANS models are also reported in the literature for their accuracy in replicating manhole hydraulics properly and efficiently [29,34,40]. The model uses a single set of Navier–Stokes equations (Equations (4) and (5)) for both fluids with additional equations to describe the free-surface (Equation (6)). The interFoam within RNG k-ε Reynolds-averaged Navier–Stokes equations also requires Equations (7) and (8).

$$\nabla \cdot u = 0 \qquad (4)$$

$$\frac{\partial \rho u}{\partial t} + \nabla \cdot (\rho u u) = -\nabla p^* + \nabla \cdot \tau - g \cdot x \nabla \rho + f_\sigma \qquad (5)$$

$$\frac{\partial \alpha}{\partial t} + \nabla \cdot (\alpha u) + \nabla \cdot [u_c \alpha (1-\alpha)] = 0 \qquad (6)$$

$$\frac{\partial \rho k}{\partial t} + \nabla \cdot (\rho k u) = \nabla \cdot (\Gamma_k \nabla k) + P_k - Y_k \qquad (7)$$

$$\frac{\partial \rho \varepsilon}{\partial t} + \nabla \cdot (\rho \varepsilon \boldsymbol{u}) = \nabla \cdot (\Gamma_\varepsilon \nabla \varepsilon) + P_\varepsilon - Y_\varepsilon + D_\varepsilon \qquad (8)$$

Where u is the mean velocity vector in the Cartesian coordinate; ρ is the density of the fluid mix; g is the acceleration due to gravity; t is the time; τ is the shear stress tensor; p^* is the modified pressure adapted by removing the hydrostatic pressure from the total pressure; f_σ is the volumetric surface tension force (where CSF and interface curvature are included); α is the VOF function; k is the turbulent kinetic energy; ε is the energy dissipation; Γ_k and Γ_ε are the diffusion for k and ε, respectively; and P, Y and D are the Production, Dissipation and Additional term for RNG, respectively.

In this work, an additional solute transport model was added to the interFoam VOF model. The main advection–dispersion equation used in the model is

$$\frac{\partial c}{\partial t} + \nabla \cdot \left(\boldsymbol{u} - v_s \frac{\boldsymbol{g}}{|\boldsymbol{g}|} \right) c = \nabla \cdot (\alpha v_t \nabla c) \qquad (9)$$

where c is the solute concentration of the flow, v_s is the terminal velocity due to gravity (which is zero for a neutrally buoyant solute) and v_t is the turbulent kinematic viscosity of water, which is a function of the turbulence of the flow [41] and taken to be equivalent to diffusivity [42,43]. The multiplication of v_t by α prevents solute particles from entering the air phase [42].

Earlier model validation works by the authors presenting measured velocities using PIV within the same experimental facility [29] showed that RNG k-ε model is a suitable RANS modelling choice for predicting water elevation and velocity profiles and hence is chosen for this work. This turbulence model can also capture complex flow and is known for better performance for separating flow [22,23,29]. Apart from wall boundary condition, five open boundaries were prescribed in the model: two inlet and two outlet boundaries at the sewer pipe and floodplain, respectively, and an atmosphere boundary at the floodplain (Figure 1). The inlet boundaries were prescribed as fixed velocities, while the outlets were applied as fixed pressure. This measured temporal mean pressure data was used for the sewer outlet pressure boundary condition (measured at P_{OutS}). The atmosphere boundary was set as equal to atmospheric pressure and zero gradient for velocity to have free airflow if required. All the wall boundaries were prescribed as noSlip condition. The sewer pipe walls were considered as rough wall applying equivalent sand roughness height (ks). Further details of measured head losses within the experimental facility can be found in [26].

Cfmesh v1.1 [44] was used to generate the hexahedral computational meshes, keeping the maximum mesh size as 10 mm towards all three Cartesian directions. The boundary meshes were kept small in such a way that $30 < y^+ < 300$, keeping three boundary layers at the all wall boundaries. A standard wall function was applied to all the walls, which has been shown to be appropriate for the application of boundary turbulence effects for such mesh sizes [36], eliminating the necessity of fine layered boundary meshes. Figure 2 shows part of the computational mesh created for this work. The rest of the CFD model such as the choice of different meshes, different solvers parameters and solution schemes were obtained from another CFD model validated in an earlier work depicting the same experimental set-up [23,29]. The maximum Courant–Friedrichs–Lewy (CFL) number was kept as 0.9. The cluster computing system at the University of Coimbra was used to run the simulations using MPI mode. Each simulation was run for 300 s to reach steady state conditions. For comparisons with experimental datasets, the measured solute concentration for each test condition was applied through the sewer inlet pipe at Cyclopes 1 when the hydraulic model reached a steady state. Unsteady model results were saved at every 0.01 s interval. Model solute concentrations were extracted at different sections and compared with the experimental measurements.

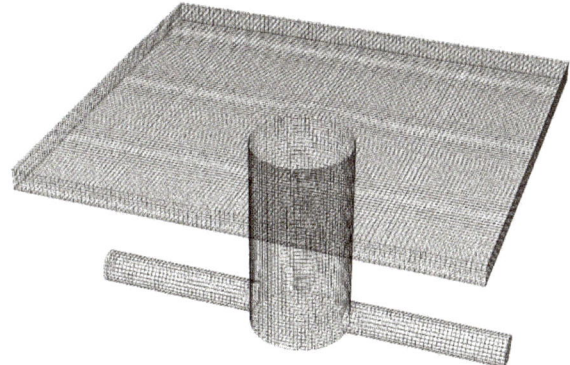

Figure 2. Computational mesh for the study showing the manhole and its connections to sewer pipes and surface.

Experimental tests 1–6 (including repeats) listed in Table 1 were replicated to perform calibration and testing of the CFD model. In these tests, solute concentration in the inflow pipe was taken as the measured value recorded at Cyclopes 1. Measured and predicted concentration curves are compared at the downstream measurement point (Cyclopes 2) along with measured and modelled flow rates in the pipe and exchanging to the surface (i.e., flow partition).

To isolate and understand the effects of the manhole separately from the pipe network, the calibrated model was further applied to Test 1-A, 2-A, 3-A and 4-A under identical hydraulic conditions to tests 1, 2, 3 and 4, respectively. However, in these tests, solute concentration was applied uniformly at the sewer pipe/manhole inlet boundary at a steady concentration of 1.2×10^{-6} mg/L. Resulting concentration time series were extracted at the manhole–pipe junctions (section A and B in Figure 1) and at the exit to the surface flow (section C in Figure 1). Solute mass flow rates to the surface flow and the downstream pipe as well as characteristic time scales of the manhole as described in Section 2.1 were calculated from these tests (described further in Section 3.3). Test 1 was further extended numerically by changing the downstream boundary pressure; by decreasing 9.5% (Test 1-B) and by increasing 15% (Test 1-C) to enable lower and higher pipe to surface exchange flows (1.21 L/s and 2.43 L/s), respectively.

3. Results

3.1. Model Calibration and Validation—Hydrodynamics

Calibration of wall roughness (ks) was performed based on experimental test results of flow exchange through the manhole to the surface. Applying a higher ks in the sewer pipe leads to lower flow through the outlet pipe with higher flow exchange from the manhole to the surface, and vice versa. The experimental values from Test 4 were used for calibration purposes as it had a sewer inlet flow which was median to all the sewer flows tested herein. Modelled ks values ranging from 1×10^{-6} to 1×10^{-3} mm were simulated in the CFD model. Results showed that ks = 0.0005 mm gives a comparable modelled value of the flow partition to the experimentally observed values (Q_e within 1.7%). This value of ks is valid for smooth surfaces such as acrylic which is appropriate to the experimental setup used here. The same ks value was applied to the rest of the hydraulic simulations (Tests 1–3 and 5–6) for model validation. Table 1 compares experimentally measured and modelled steady state flow rates in the pipe and exchanged to the surface (Q_{OutS} and Q_e) for each test, along with measured boundary conditions and calculated Reynolds numbers. Modelled and measured flow rates are found to be within 1.7% in all cases. Figure 3 presents resulting calculated velocity streamlines and vectors within the manhole during Test 4.

Table 1. Experimentally observed and numerical flow rates for each test case. Solute injections for tests 4–6 were repeated 3 times.

Test ID	Boundary Condition			Experimental		Numerical		Experimental Reynold's No.		% of Diff. in Q_e
	U/S Q_{InF}(L/s)	U/S Q_{InS}(L/s)	D/S P_{outS}(mm)	Q_{OutS} (L/s)	Q_e(L/s)	Q_{OutS} (L/s)	Q_e(L/s)	Inlet Sewer	Outlet Sewer	
Test 1	4.28	8.09	415.9	6.42	1.67	6.44	1.65	137020	108680	1.20
Test 2	4.28	9.00	428.3	6.84	2.17	6.84	2.16	152480	115790	0.46
Test 3	4.28	9.67	436.7	7.18	2.49	7.14	2.53	163830	121620	1.61
Test 4 *	6.29	10.20	448.7	6.72	3.48	6.66	3.54	172710	113830	1.72
Test 5	7.46	10.20	450.2	6.70	3.50	6.65	3.55	172710	113490	1.43
Test 6	8.64	10.19	447.5	6.67	3.52	6.66	3.53	172710	112980	0.28

* Test 4 data was used for hydraulic calibration.

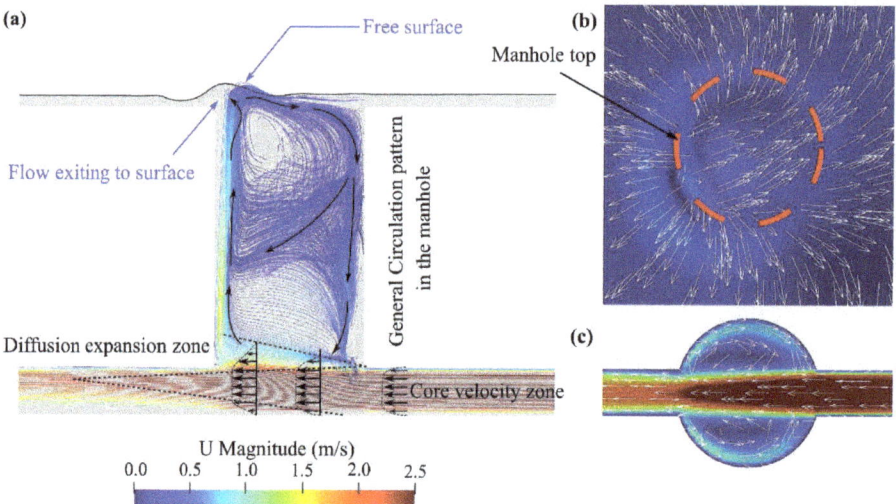

Figure 3. Hydraulic conditions inside the manhole during Test 4. (a) Streamline of the flow indicating a general circulation pattern, (b) mean velocity vectors at the top horizontal plane of the manhole and (c) mean velocity vectors at the horizontal plane passing through the sewer pipe axis of the manhole. At all cases, main flow direction is from right to the left.

3.2. Model Replication of Mixing Processes within the Manhole

Following validation of the hydrodynamic processes, the ability of the CFD model to simulate solute mixing within the manhole was tested by comparing measured and simulated concentration profiles within the pipe network (at the location of Cyclopes 2) for all hydraulic conditions. Solute injections for hydraulic conditions in Tests 1–3 were performed once using either a single or double pulse of solute concentration. Injections during hydraulic conditions in Tests 4–6 were repeated three times each, of which the first two had single pulse and the third had two consecutive concentration pulses. Measured and predicted solute concentration time series at manhole D/S (at the location of Cyclopes 2) were extracted compared to those of experimental data. Figure 4 shows comparison of experimentally measured and modelled concentration time series at the manhole downstream measurement point for each test.

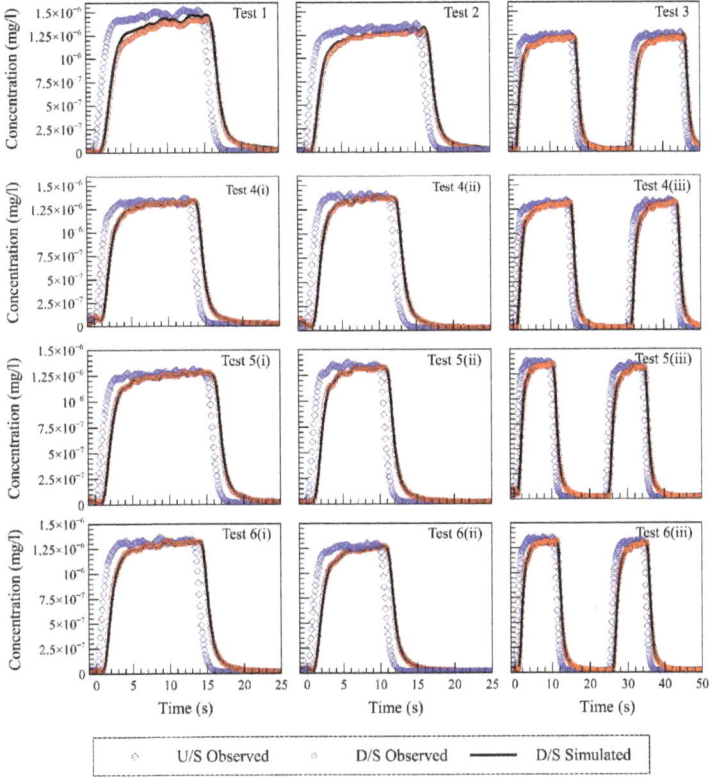

Figure 4. Comparison of experimental and numerical unsteady concentration profiles in the sewer pipe downstream of the manhole (Cyclopes 2) and the measured concentration at the upstream of the manhole (Cyclopes 1). Tests 4–6 are repeated three times (i, ii, iii).

Different statistical parameters were used to check the quality of model performance in predicting the solute concentration at the downstream of the manhole. The parameters used are listed below.

- Average of error, $BIAS = \frac{1}{n} \sum_{i=1}^{n} (O_i - P_i)$

- Root mean square error, $RMSE = \frac{1}{n} \sqrt{\sum_{i=1}^{n} (O_i - P_i)^2}$

- Pearson product-moment correlation coefficient, $r = \frac{\sum_{i=1}^{n} (O_i - \overline{O})(P_i - \overline{P})}{\sqrt{\sum_{i=1}^{n} (O_i - \overline{O})^2 (P_i - \overline{P})^2}}$

- Nash Sutcliffe coefficient, $NSC = 1 - \frac{\sum_{i=1}^{n} (O_i - P_i)^2}{\sum_{i=1}^{n} (O_i - \overline{O})^2}$

Where O_i is the observation values, P_i is the model predicted values, and $\overline{O}$ is the average of all observed values, $\overline{P}$ is the average of model predicted values and n is the number of observations. The calculated values of the mentioned statistical parameters are shown in Table 2. It shows that BIAS of all the comparisons is negligible. The NSC values are greater than 0.995 in all cases. Therefore, the results show that the model accurately replicates the solute mixing processes within the manhole.

Table 2. Statistical comparisons between the concentration time series between experimental and numerical models.

Test ID	BIAS (mg/L)	RMSE (mg/L)	r	NSC
Test 1	-3.89×10^{-8}	4.46×10^{-8}	1.000	0.995
Test 2	-2.48×10^{-8}	2.78×10^{-8}	1.000	0.997
Test 3	-1.39×10^{-8}	1.87×10^{-8}	1.000	0.999
Test 4 (i)	3.41×10^{-9}	1.45×10^{-8}	1.000	0.999
Test 4 (ii)	2.53×10^{-9}	1.27×10^{-8}	1.000	1.000
Test 4 (iii)	4.70×10^{-9}	1.90×10^{-8}	0.999	0.999
Test 5 (i)	3.68×10^{-9}	2.28×10^{-8}	1.000	0.999
Test 5 (ii)	3.62×10^{-9}	2.69×10^{-8}	1.000	0.998
Test 5 (iii)	4.55×10^{-9}	1.25×10^{-8}	1.000	1.000
Test 6 (i)	5.01×10^{-9}	2.59×10^{-8}	1.000	0.999
Test 6 (ii)	3.03×10^{-9}	2.62×10^{-8}	1.000	0.998
Test 6 (iii)	6.57×10^{-9}	2.97×10^{-8}	0.999	0.997

3.3. Modelling of Soluble Mass Exchange to Surface Flows

The solute transport model was then applied to Test 1-A-B-C, 2-A, 3-A and 4-A, as described in Section 2.3 (i.e., with a uniform solute applied directly to the manhole inlet boundary at Section A). Figure 5 shows example plots of concentration evolution inside the manhole for each of these tests at different time intervals. Time $t_0 = 0$ is taken when average solute concentration at Section A exceeds 1% of the peak value. Instantaneous velocity vectors are also displayed to indicate the travel paths of the solute concentration within the manhole volume.

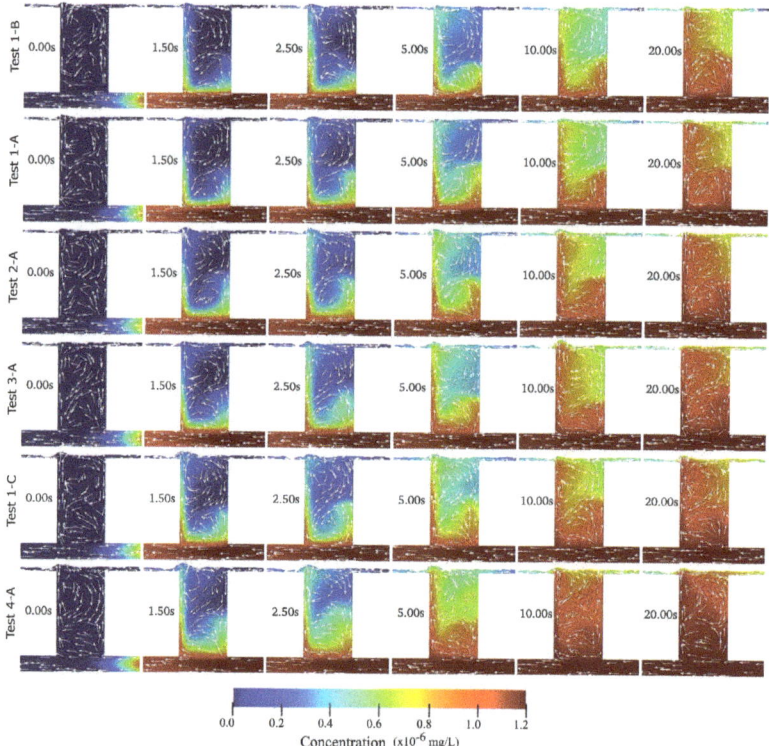

Figure 5. Instantaneous solute concentrations and velocity vectors at the central vertical plane of the manhole from different tests at time (t) from 0–20 s. The arrows are not drawn to scale.

Figure 5 shows that as soon as the solute mass enters the manhole, it diffuses from the high velocity flow region into the manhole volume. A part of the concentrated solute mass hits the opposite manhole wall and travels towards the manhole surface. Later, it interacts with the surface flow and recirculates to the manhole. This recirculating flow brings low concentration flow from the surface into the manhole, maintaining a consistent concentration gradient through the manhole height until the upper part becomes completely mixed. The observed flow structures explored in the tests are relativity insensitive to the pipe inflow rate over the partition ratios used in these tests. The results show that until well mixed conditions are achieved, that the concentration field at the manhole/surface interaction point (section C) is highly heterogeneous. Therefore unlike in the pipe network, (where cross-sectionally averaged values can be reasonably assumed at Sections A and B), quantification of mass flow rate to the surface (i.e., over Section C) requires robust understanding of the spatial variation of solute and velocity over the manhole cross section and how this evolves with time.

The evolution of solute mass exchange through each cross section A, B and C is quantified based on the CFD model. For this purpose, CFD model results of test 1-A-B-C, 2-A, 3-A and 4-A were considered. Due to highly heterogeneous conditions at section C, mass flow rate at each time step for each inlet/outlet junctions (section A, B, C) was calculated using the following Equation,

$$\dot{M}_x = \int_{i=0}^{i=A} c_i u_i dA \quad (10)$$

where $\dot{M}_x$ is the solute mass flow rate though section A, B or C (i.e., $\dot{M}_{PI}, \dot{M}_e$ or $\dot{M}_{PO}$); u_i is the mean velocity vector normal to area i; dA is an incremental cross section area vector (based on a 10mm slice); and c_i is the solute concentration within area i. Hence the integral value of the dot multiplication of these components is used to provide the net mass flow rate through sections A, B and C. The model set-up (uniform concentration applied at Section A) results in a constant $\dot{M}_{PI}$ over each test after the first 0.2 s of simulation (as given in Table 3). Following the calculation of $\dot{M}_{PI}, \dot{M}_e$ and $\dot{M}_{PO}$, the rate of change in solute mass within the manhole was calculated using Equation (1). Figure 6 shows resulting outlet solute mass flow rates as a ratio of manhole inlet mass flow rate over each test. The time axis in the figures represents time (in seconds) since the first solute enters the manhole from the sewer inlet. As in Figure 5, this time ($t_0 = 0$) is taken when average solute concentration at Section A exceeds 1% of the peak value. Significant fluctuation can be observed in the $\frac{\dot{M}_e}{\dot{M}_{PI}}$ values due to the complex heterogeneous nature of the flow at the surface/manhole interaction point (section C).

Table 3. Characteristic time scales of solute mixing from different model results. Results are arranged in an ascending order of the mean surface flow partition ratios.

Test ID	Inlet Mass Flowrate, $\dot{M}_{PI}$ (×10^{-6} mg/s)	Nominal Residence Time, T_x (s)	Mean Surface Flow Partition Ratio (Q_e/Q_{inS})	Non-Dimensional Characteristic Time (-)			Fitted Curve Coefficients	
				t_1/T_x	t_2/T_x	t_3/T_x	C	r^2
1-B	9.7	4.09	0.150	0.03	0.20	5.80	0.35	0.9071
1-A	9.7	2.96	0.206	0.04	0.27	8.11	0.40	0.9677
2-A	10.8	2.28	0.241	0.05	0.34	8.27	0.40	0.9455
3-A	11.6	1.99	0.257	0.05	0.25	9.20	0.45	0.9545
1-C	9.7	2.24	0.302	0.05	0.26	9.79	0.52	0.9797
4-A	12.3	1.42	0.342	0.06	0.28	9.55	0.60	0.9356

Characteristic time scales, as described in Section 2.2, are defined for each test and are presented in Table 3. Similar to the definition of t_0, t_1 and t_2 are taken when averaged solute concentration at Sections B and C exceeds 1% of the peak value, respectively. t_3 is defined as the time when $\frac{dM_m}{dt}$ falls below 2.5% of its peak value for the first time. Timescales in Table 3 are presented non-dimensionally in terms of the nominal manhole residence time for flow passing to the surface T_x, as calculated using Equation (11),

$$T_x = \frac{L_x \frac{\pi}{4}(\Phi_m)^2}{Q_e} \qquad (11)$$

where L_x is the vertical distance between the sewer pipe axis and the manhole top and Φ_m is the manhole cross sectional area.

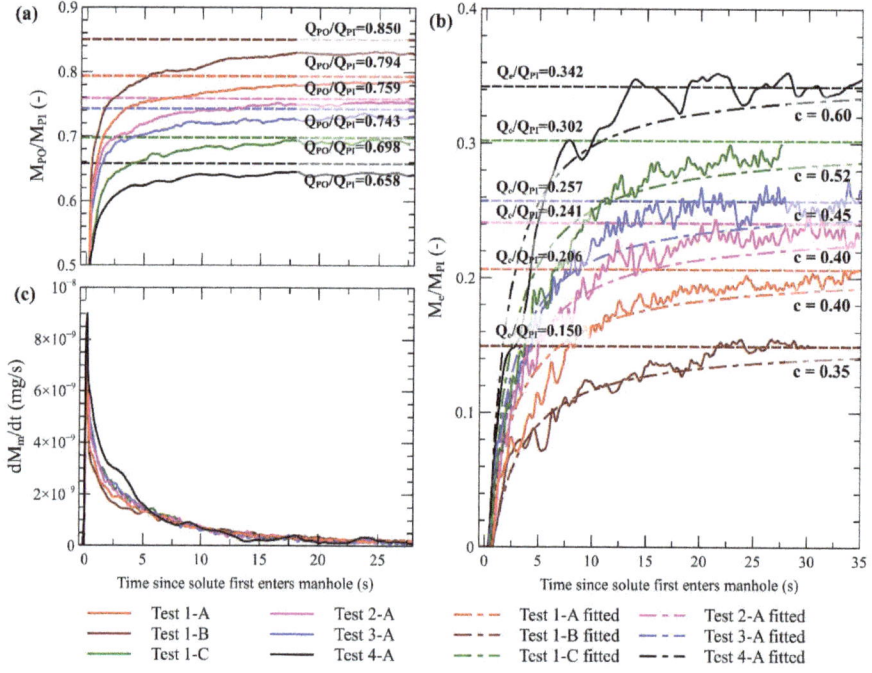

Figure 6. (a) Mass exchange ratio at the manhole to sewer pipe outlet from different test results. Horizontal lines indicate Q_{PO}/Q_{inS} values for each test. (b) Mass exchange ratio at the manhole to surface connection with fitted asymptotic trend lines. Horizontal lines indicate Q_e/Q_{inS} values for each test. (c) Change in solute mass within the manhole.

As can be seen in Figure 6a,b, for each test the proportion of mass flow rate entering the pipe outlet (M_{PO}/M_{PI}) and surface flow (M_e/M_{PI}) grows asymptotically toward the relevant flow partition ratio (as defined in Equations (2) and (3)). Therefore, solute mass flow exchange to the surface can be described using the following function,

$$\frac{M_e}{M_{Pin}} = \left(\frac{-1}{C(t-t_2)+1} + 1\right)\frac{Q_e}{Q_{inS}} \qquad (12)$$

where C is an empirical coefficient. The best fit value of C and resulting goodness of fit (r^2) value between fitted equation and CFD model results for each test conducted in this work is given in Table 3. The first arrival of mass at the pipe exit (t_1) and the surface flow (t_2) occurs relatively quickly in all conditions (0.09 s < t_1 < 0.13 s and 0.40 s < t_2 < 0.82 s), while the timescale for complete mixing (t_3) to be achieved (and thus mass flow rate to the surface flow to become approximately steady and equivalent to the surface flow partition ratio) varies significantly over the tests conducted (13.6 s < t_3 < 24.0 s). From

Table 3 the value of C and the non-dimensional timescale ($t_{1,2,3}/T_x$) to achieve well mixed conditions tend to increase with the flow partition ratio over the range of conditions tested.

4. Discussion

A comparison of experimentally measured and modelled discharges within a scaled manhole structure shows that, given knowledge of the boundary conditions, the RANS CFD approach accurately simulates flow exchange from piped to surface flows (within 1.7% in all test cases). Therefore, steady-state flow exchange through similar hydraulic structures during flood events is likely to be well described using RANS CFD. These results concur with previous validation studies utilising similar 3D modelling approaches to simulate hydrodynamics in urban drainage structures (see, e.g., in [29,45]), although in this case the complex interaction with surface flows as well as a solute transport is also recreated. Such models are too computationally expensive to be used in direct flood modelling applications; however, there is further potential to utilise such complex models to evaluate simpler semiempirical weir/orifice relationships currently used to describe surface/sewer interaction. Such semiempirical relationships have been found to be sensitive to interaction structure type, inlet characteristics and geometry as well as unsteady hydraulic conditions [12,30,46,47], and thus benefit from case-specific calibration. Similarly, the calibrated model has been shown to accurately reproduce solute concentration profiles (and thus mass flow rates) measured downstream of the manhole structure under a range of flow rates during cases where sewer flow interacts with surface flood water. Taken together with the agreement of modelled and measured flow partition within the manhole, as well as past results comparing CFD velocity vectors against those obtained using PIV measurement in the same facility [29], this result gives confidence that the CFD model can reproduce flow details and resulting solute mass exchange to the surface during flood conditions. A full validation would benefit from having access to measured values of concentration/mass exchange at the interaction point between sewer and surface flows (section C); however, the current results have demonstrated the hydraulic complexity and spatial heterogeneity of concentration at this position. Therefore, such validation measurements would require complex instrumentation such as Laser-Induced Fluorescence (LIF) to provide detailed spatial data over the manhole area. In addition, further validation of CFD approaches would be valuable in more complex hydraulic conditions (e.g., unsteady flow), in systems with different geometrical features or at different scales or in cases involving sediments which are also commonly present in urban drainage networks and may be susceptible to transportation in flood water.

The modelled flow structures illustrate the complexity of the interaction between surcharging manhole flow and surface flood water; however, flow structures within the manhole appear to be relativity insensitive to the pipe inflow rate over the flow partition ratios explored in these tests. The solute transport and resulting mass flow rates within the system are a process of both advection and diffusion. The solute transport from the manhole inlet to the manhole pipe outlet is dominated by the advection process due to the strong local velocities in this zone. Thus, first arrival time to the sewer outlet (t_1) is dominated by the sewer inlet velocity with little subsequent variation within these tests. In addition, as the flow partition ratio increases (i.e., more flow is transported to the surface) the corresponding timescales for first arrival of mass at the surface (t_2) and complete mixing within the manhole (t_3) decrease slightly due to the increasing advection through the manhole structure to the surface. However, a stronger positive relationship is observed between non-dimensional timescales based on the characteristic manhole residence time (t_2/T_x, t_3/T_x) and surface partition ratio (Q_e/Q_{inS}), indicating the relative significance of conical flow structures produced by the inlet pipe and subsequent diffusive mixing processes in the tests conducted.

The work has shown that the sharp arrival of a well-mixed solute at an open manhole results in an asymptotic growth of mass exchange to the surface, converging to a value that is defined by the hydraulic flow partition. Parameterisation of an asymptotic growth function (C) may be related to the flow partition ratio and/or the characteristic residence time, with more rapid mixing occurring at lower residence times. Approximately well-mixed conditions (and associated equivalence of sewer to surface solute mass exchange s and flow partition ratios) occur at between 5.8 and 9.2 times the manhole residence time over the conditions tested here. Further work is required to explore these relationships over a range of manhole geometries, using different (time varying) solute injection profiles and unsteady hydraulic conditions as well as in other exchange structures such as gullies featuring grills/covers, including at full scale, such that realistic timescales in real situations can be established. A more complete understanding of this problem should also consider the transport of solids, such as fine sediments and entrained material, which are also present in urban drainage networks. In addition, other flood scenarios (e.g., further exploring the influence of surface flow depth and velocity) and cases where the majority of flow transfers to the surface ($Q_e/Q_{inS} > 0.5$) could be explored. In such cases where the surface flow partition ratio is significantly larger, the bulk advection of solute by the flow is likely to increasingly dominate diffusivity arising from local flow structures.

5. Conclusions

A 3D CFD model was applied to simulate flows in an exchange structure involving interacting pipe and surface flows to quantify flow and soluble pollutant mass exchange. The model was validated with a laboratory-scale model, achieving differences of less than 1.7% in flow rates and excellent statistical comparisons between observed and modelled concentration time series. This suggests that a RANS CFD approach is an appropriate methodology to evaluate flow partition and to evaluate how soluble pollutants move from sewer networks to surface flood flows.

The model was extended to different conditions to understand the effects of the manhole separately from the pipe network, and used to calculate the evolution of solute mass transport rate through each manhole open boundary cross section under a range of flow conditions including interactions between sewer flows and surface flood water. A sharp arrival of solute into the structure is shown to result in an asymptotic growth of solute mass exchange ratio to the surface converging to a value equal to the surface flow partition ratio. An analysis of the results demonstrates that the timescales to achieve this convergence are dependent on the diffusive transport inside the structure.

The work in this paper describes initial steps to understand the risks of soluble material from sewer networks entering urban flood waters via exchange structures. The transport of pollutants through these structures will also depend on additional factors including, but not limited to, the presence of manhole coverings and change of structure geometry/shape. In order to build a more complete understanding, such that risks to public health can be understood and quantified, requires significantly more work. This includes further consideration of transport and transformations of both contaminated sediments and soluble materials in urban drainage networks as well as datasets from urban drainage networks, floodwaters and urban surfaces such that transport, survival and fate can be modelled within quantifiable uncertainty bounds.

Author Contributions: Conceptualization, R.F.C., J.D.S. and M.R.; methodology, M.N.A.B. and M.R.; software, M.N.A.B. and R.F.C.; validation, M.N.A.B.; formal analysis, M.N.A.B.; writing—original draft preparation, M.N.B. and J.S.; writing—review and editing, J.D.S., M.R. and R.F.C.; visualization, M.N.A.B.; supervision, J.D.S. and R.F.C. All authors have read and agreed to the published version of the manuscript.

Funding: The research was supported by the UK Engineering and Physical Sciences Research Council (EP/K040405/1). Author Beg MNA worked as part of QUICS (Quantifying Uncertainty in Integrated Catchment Studies) project which received funding from the European Union's Seventh Framework Programme (Grant agreement No. 607000). Cluster computing system used in this study is supported by the Laboratory for Advanced Computing of the University of Coimbra. Authors would also like acknowledge the support of FCT (Portuguese Foundation for Science and Technology) through the Project UIDB/04292/2020-MARE.

Conflicts of Interest: The authors declare no conflict of interest.

Data Available: Additional datasets associated with this work can be downloaded from https://zenodo.org/communities/floodinteract/.

References

1. Arnell, N.W.; Gosling, S.N. The impacts of climate change on river flood risk at the global scale. *Clim. Chang.* **2016**, *134*, 387–401. [CrossRef]
2. Acquaotta, F.; Faccini, F.; Fratianni, S.; Paliaga, G.; Sacchini, A.; Vilímek, V. Increased flash flooding in Genoa Metropolitan Area: A combination of climate changes and soil consumption? *Meteorol. Atmos. Phys.* **2019**, *131*, 1099–1110. [CrossRef]
3. Rubinato, M.; Nichols, A.; Peng, Y.; Zhang, J.-m.; Lashford, C.; Cai, Y.-p.; Lin, P.-z.; Tait, S. Urban and river flooding: Comparison of flood risk management approaches in the UK and China and an assessment of future knowledge needs. *Water Sci. Eng.* **2019**, *12*, 274–283. [CrossRef]
4. Simões, N.E.; Ochoa-Rodríguez, S.; Wang, L.P.; Pina, R.D.; Marques, A.S.; Onof, C.; Leitão, J.P. Stochastic urban pluvial flood hazard maps based upon a spatial-temporal rainfall generator. *Water* **2015**, *7*, 3396–3406. [CrossRef]
5. Wan Mohtar, W.H.M.; Abdullah, J.; Abdul Maulud, K.N.; Muhammad, N.S. Urban flash flood index based on historical rainfall events. *Sustain. Cities Soc.* **2020**, *56*. [CrossRef]
6. Mort, M.; Walker, M.; Williams, A.L.; Bingley, A. Displacement: Critical insights from flood-affected children. *Health Place* **2018**, *52*, 148–154. [CrossRef]
7. De Man, H.; Van Den Berg, H.H.J.L.; Leenen, E.J.T.M.; Schijven, J.F.; Schets, F.M.; Van Der Vliet, J.C.; Van Knapen, F.; De Roda Husman, A.M. Quantitative assessment of infection risk from exposure to waterborne pathogens in urban floodwater. *Water Res.* **2014**, *48*, 90–99. [CrossRef]
8. Sales-Ortells, H.; Medema, G. Microbial health risks associated with exposure to stormwater in a water plaza. *Water Res.* **2015**, *74*, 34–46. [CrossRef]
9. Ten Veldhuis, J.A.E.; Clemens, F.H.L.R.; Sterk, G.; Berends, B.R. Microbial risks associated with exposure to pathogens in contaminated urban flood water. *Water Res.* **2010**, *44*, 2910–2918. [CrossRef]
10. Collender, P.A.; Cooke, O.C.; Bryant, L.D.; Kjeldsen, T.R.; Remais, J.V. Estimating the microbiological risks associated with inland flood events: Bridging theory and models of pathogen transport. *Crit. Rev. Environ. Sci. Technol.* **2016**, *46*, 1787–1833. [CrossRef]
11. Rubinato, M.; Martins, R.; Kesserwani, G.; Leandro, J.; Djordjević, S.; Shucksmith, J. Experimental calibration and validation of sewer/surface flow exchange equations in steady and unsteady flow conditions. *J. Hydrol.* **2017**, *552*, 421–432. [CrossRef]
12. Beg, M.N.A.; Carvalho, R.F.; Leandro, J. Effect of surcharge on gully-manhole flow. *J. Hydro-Environ. Res.* **2018**, *19*, 224–236. [CrossRef]
13. Carvalho, R.F.; Lopes, P.; Leandro, J.; David, L.M. Numerical research of flows into gullies with different outlet locations. *Water* **2019**, *11*, 794. [CrossRef]
14. Hammond, M.J.; Chen, A.S.; Djordjević, S.; Butler, D.; Mark, O. Urban flood impact assessment: A state-of-the-art review. *Urban Water J.* **2015**, *12*, 14–29. [CrossRef]
15. Mark, O.; Jørgensen, C.; Hammond, M.; Khan, D.; Tjener, R.; Erichsen, A.; Helwigh, B. *A New Methodology for Modelling of Health Risk from Urban Flooding Exemplified by Cholera—Case Dhaka, Bangladesh*; Blackwell Publishing Inc.: Hoboken, NJ, USA, 2018; Volume 11, pp. 28–42.
16. Tscheikner-Gratl, F.; Bellos, V.; Schellart, A.; Moreno-Rodenas, A.; Muthusamy, M.; Langeveld, J.; Clemens, F.; Benedetti, L.; Rico-Ramirez, M.A.; de Carvalho, R.F.; et al. Recent insights on uncertainties present in integrated catchment water quality modelling. *Water Res.* **2019**, *150*, 368–379. [CrossRef]
17. Scoullos, I.M.; Adhikari, S.; Lopez Vazquez, C.M.; van de Vossenberg, J.; Brdjanovic, D. Inactivation of indicator organisms on different surfaces after urban floods. *Sci. Total Environ.* **2020**, *704*. [CrossRef]
18. Schreiber, C.; Heinkel, S.B.; Zacharias, N.; Mertens, F.M.; Christoffels, E.; Gayer, U.; Koch, C.; Kistemann, T. Infectious rain? Evaluation of human pathogen concentrations in stormwater in separate sewer systems. *Water Sci. Technol.* **2019**, *80*, 1022–1030. [CrossRef]
19. Martins, R.; Kesserwani, G.; Rubinato, M.; Lee, S.; Leandro, J.; Djordjević, S.; Shucksmith, J.D. Validation of 2D shock capturing flood models around a surcharging manhole. *Urban Water J.* **2017**, *14*, 892–899. [CrossRef]
20. Guymer, I.; Dennis, P.; O'Brien, R.; Saiyudthong, C. Diameter and Surcharge Effects on Solute Transport across Surcharged Manholes. *J. Hydraul. Eng.* **2005**, *131*, 312–321. [CrossRef]
21. Guymer, I.; Stovin, V.R. One-Dimensional Mixing Model for Surcharged Manholes. *J. Hydraul. Eng.* **2011**, *137*, 1160–1172. [CrossRef]

22. Lau, S.D.; Stovin, V.R.; Guymer, I. The prediction of solute transport in surcharged manholes using CFD. *Water Sci. Technol.* **2007**, *55*, 57–64. [CrossRef] [PubMed]
23. Beg, M.N.A.; Carvalho, R.F.; Leandro, J. Effect of manhole molds and inlet alignment on the hydraulics of circular manhole at changing surcharge. *Urban Water J.* **2019**, *16*, 33–44. [CrossRef]
24. Stovin, V.R.; Guymer, I.; Lau, S.-T.D. Dimensionless method to characterize the mixing effects of surcharged manholes. *J. Hydraul. Eng.* **2010**, *136*, 318–327. [CrossRef]
25. Beg, M.N.A. Detailed Uncertainty Analysis of Urban Hydraulic Structures in Large Catchments. Ph.D. Thesis, Department of Civil Engineering, University of Coimbra, Coimbra, Portugal, 2018.
26. Rubinato, M. Physical scale modelling of urban flood systems. Ph.D. Thesis, Department of Civil and Structural Engineering, University of Sheffield, Sheffield, UK, 2015.
27. Rubinato, M.; Martins, R.; Shucksmith, J.D. Quantification of energy losses at a surcharging manhole. *Urban Water J.* **2018**, *9006*, 1–8. [CrossRef]
28. Rubinato, M.; Shucksmith, J.; Saul, A.J.; Shepherd, W. Comparison between InfoWorks hydraulic results and a physical model of an Urban drainage system. *Water Sci. Technol.* **2013**, *68*, 372–379. [CrossRef]
29. Beg, M.N.A.; Carvalho, R.F.; Tait, S.; Brevis, W.; Rubinato, M.; Schellart, A.; Leandro, J. A comparative study of manhole hydraulics using stereoscopic PIV and different RANS models. *Water Sci. Technol.* **2018**, *2017*, 87–98. [CrossRef]
30. Rubinato, M.; Lee, S.; Martins, R.; Shucksmith, J.D. Surface to sewer flow exchange through circular inlets during urban flood conditions. *J. Hydroinform.* **2018**, *20*, 564–576. [CrossRef]
31. Martins, R.; Rubinato, M.; Kesserwani, G.; Leandro, J.; Djordjević, S.; Shucksmith, J.D. On the Characteristics of Velocities Fields in the Vicinity of Manhole Inlet Grates During Flood Events. *Water Resour. Res.* **2018**, *54*, 6408–6422. [CrossRef]
32. Defra. *Annex B: National Build. Standards Design and Construction of New Gravity Foul Sewers and Lateral Drains Water Industry Act. 1991 Section 106B Flood and Water Management Act. 2010 Section 42*; Defra: London, UK, 2011.
33. Gotfredsen, E.; Kunoy, J.D.; Mayer, S.; Meyer, K.E. Experimental validation of RANS and DES modelling of pipe flow mixing. *Heat Mass Transf.* **2020**, *56*, 2211–2224. [CrossRef]
34. Stovin, V.R.; Bennett, P.; Guymer, I. Absence of a Hydraulic Threshold in Small-Diameter Surcharged Manholes. *ASCE J. Hydraul. Eng.* **2013**, *139*, 984–994. [CrossRef]
35. Mark, O.; Ilesanmi-Jimoh, M. An analytical model for solute mixing in surcharged manholes. *Urban Water J.* **2017**, *14*, 443–451. [CrossRef]
36. Weller, H.G.; Tabor, G.; Jasak, H.; Fureby, C. A tensorial approach to computational continuum mechanics using object-oriented techniques. *Comput. Phys.* **1998**, *12*, 620. [CrossRef]
37. Jasak, H. Error Analysis and Estimation for the Finite Volume Method with Applications to Fluid Flows. Ph.D. Thesis, University of London, London, UK, 1996.
38. Rusche, H. Computational Fluid Dynamics of Dispersed Two-Phase Flows at High Phase Fractions. Ph.D. Thesis, University of London, London, UK, 2002.
39. Hirt, C.W.; Nichols, B.D. Volume of fluid (VOF) method for the dynamics of free boundaries. *J. Comput. Phys.* **1981**, *39*, 201–225. [CrossRef]
40. Bennett, P. Evaluation of the Solute Transport Characteristics of Surcharged Manholes using a RANS Solution. Ph.D. Thesis, Department of Civil and Structural Engineering, University of Sheffield, Sheffield, UK, 2012.
41. Yakhot, V.; Orszag, S.A.; Thangam, S.; Gatski, T.B.; Speziale, C.G. Development of turbulence models for shear flows by a double expansion technique. *Phys. Fluids* **1992**, *4*, 1510–1520. [CrossRef]
42. Jacobsen, N.G. A Full Hydro and Morphodynamic Description of Breaker Bar Development. Ph.D. Thesis, Technical University of Denmark, Lyngby, Denmark, 2011.
43. Liu, X.; García, M.H. Three-Dimensional Numerical Model with Free Water Surface and Mesh Deformation for Local Sediment Scour. *J. Waterw. Port Coast. Ocean Eng.* **2008**, *134*, 203–217. [CrossRef]
44. Juretić, F. *cfMesh User Guide (v1.1)*; Creative Fields: Zagreb, Croatia, 2015; Volume 6.
45. Lopes, P.; Leandro, J.; Carvalho, R.F.; Páscoa, P.; Martins, R. Numerical and experimental investigation of a gully under surcharge conditions. *Urban Water J.* **2015**, *12*, 468–476. [CrossRef]

46. Gómez, M.; Russo, B.; Tellez-Alvarez, J. Experimental investigation to estimate the discharge coefficient of a grate inlet under surcharge conditions. *Urban Water J.* **2019**, *16*, 85–91. [CrossRef]
47. Kemper, S.; Schlenkhoff, A. Experimental study on the hydraulic capacity of grate inlets with supercritical surface flow conditions. *Water Sci. Technol.* **2019**, *79*, 1717–1726. [CrossRef]

© 2020 by the authors. Licensee MDPI, Basel, Switzerland. This article is an open access article distributed under the terms and conditions of the Creative Commons Attribution (CC BY) license (http://creativecommons.org/licenses/by/4.0/).

Article

Modelling Pluvial Flooding in Urban Areas Coupling the Models Iber and SWMM

Esteban Sañudo, Luis Cea * and Jerónimo Puertas

Department of Civil Engineering, Water and Environmental Engineering Group, Universidade da Coruña, Elviña, 15071 A Coruña, Spain; e.sanudo@udc.es (E.S.); jeronimo.puertas@udc.es (J.P.)
* Correspondence: luis.cea@udc.es

Received: 31 July 2020; Accepted: 18 September 2020; Published: 22 September 2020

Abstract: Dual urban drainage models allow users to simulate pluvial urban flooding by analysing the interaction between the sewer network (minor drainage system) and the overland flow (major drainage system). This work presents a free distribution dual drainage model linking the models Iber and Storm Water Management Model (SWMM), which are a 2D overland flow model and a 1D sewer network model, respectively. The linking methodology consists in a step by step calling process from Iber to a Dynamic-link Library (DLL) that contains the functions in which the SWMM code is split. The work involves the validation of the model in a simplified urban street, in a full-scale urban drainage physical model and in a real urban settlement. The three study cases have been carefully chosen to show and validate the main capabilities of the model. Therefore, the model is developed as a tool that considers the main hydrological and hydraulic processes during a rainfall event in an urban basin, allowing the user to plan, evaluate and design new or existing urban drainage systems in a realistic way.

Keywords: urban drainage; dual drainage; Iber; SWMM; urban flooding

1. Introduction

The rise of impervious areas in cities due to urbanization has increased the occurrence of flooding and its consequences during extreme rainfall events. In order to mitigate flood impacts it is essential for cities to have a drainage network properly planned and designed. Urban drainage systems are made of two clearly different subsystems: the sewer (minor) network and the surface (major) network. During a rainfall event, the water exchange between both subsystems can be in both directions through inlets and manholes. The surface overland flow, the sewer flow and the exchange between both of them are commonly computed using dual drainage models that solve all the processes in an integrated and realistic way [1–4].

The first dual drainage approaches were 1D/1D models that simplify the surface flow as open channels or ponds, solving the 1D Saint-Venant equations [5]. Modelling a street as an open channel is a sensible approach as long as the water velocities presents a preferent direction [6]. However, the flow in urban districts has a significant two-dimensional behaviour that is necessary to consider in the numerical modelling in order to achieve truthful results. Therefore, 1D/2D dual drainage models were developed, that solve the two-dimensional shallow water equations on the surface while maintaining the one-dimensional approach in the sewer network [7–9]. These 1D/2D approaches, combined with accurate digital elevation models (DEM), allow modellers to obtain more realistic results. There are different 1D/2D dual drainage models, some of them developed on research works [10], others with licensed software such as Infoworks ICM or Mike-Urban. To date, there are some free distribution 1D/2D drainage models that solve surface and sewer flow and their interaction but there are not usually used in real projects of engineering with guarantee and user-friendliness.

Storm Water Management Model (SWMM) [11] model is used frequently as 1D sewer network engine in 1D/2D urban drainage models. Thus, coupling SWMM with overland flows models solves the limitation of SWMM to simulate and visualise the flood area. There exist multiple methods of coupling SWMM with overland flows models [9,12–21] in which different coupling methodologies are used. Some models modify or adapt the SWMM code [12] while others call SWMM libraries and functions without editing the engine code [9]. Regarding the flow exchange, there are models that considerer only a unidirectional exchange from sewer network to the surface [13,15] and others that consider a bidirectional exchange that allows flow enter to the sewer [14]. The election of the overland flow model determines the complexity of the dual model and its computation time. In this way, for overland flow, there is a wide variety of models based on the equations of fully dynamic shallow water and others based on simplified models as the local inertial model, diffusive wave model or the kinematic wave [16].

In addition, there is commercial software such as XP-SWMM [20], PCSWMM [22], Mike-Urban [21] or FLO-2D Pro that couple SWMM engine with 2D overland flow models.

In this work, a 1D/2D dual urban drainage model is developed linking two free distributed hydraulic models: Iber [23] and SWMM [11]. The model links the Iber source code with the SWMM Dynamic-link Library (DLL), which contains functions that allow simulation data to be exchanged between SWMM and other software. The SWMM dynamics libraries used here were developed in [24], and up to now, they have not been implemented in any 1D/2D dual drainage model. The model presented here is therefore a new free tool that allows an accurate and complete analysis of the flood area extension and its duration due to the bidirectional exchange implemented. The validation of the model is shown in three different test cases that demonstrate its capabilities.

2. Materials and Methods

2.1. Hydraulic Models

2.1.1. Iber

Iber is a 2D numerical model for simulating turbulent free surface flow and transport processes in shallow waters [23,25]. It is a free distribution software that can be downloaded at www.iberaula.com. The hydrodynamics module of Iber solves the 2D depth-averaged shallow water equations, also known as dynamic wave equations to distinguish them from simpler diffusive and kinematic wave models. The model also incorporates several hydrological processes, and its application to rainfall-runoff and overland flow computations has been validated in previous works [26–29]. The mass and momentum conservation equations solved by the model can be written as follows:

$$\frac{\partial h}{\partial t} + \frac{\partial q_x}{\partial x} + \frac{\partial q_y}{\partial y} = R - i \tag{1}$$

$$\frac{\partial q_x}{\partial t} + \frac{\partial}{\partial x}\left(\frac{q_x^2}{h} + g\frac{h^2}{2}\right) + \frac{\partial}{\partial y}\left(\frac{q_x q_y}{h}\right) = -gh\frac{\partial z_b}{\partial x} - g\frac{n^2}{h^{7/3}}|q|q_x \tag{2}$$

$$\frac{\partial q_y}{\partial t} + \frac{\partial}{\partial x}\left(\frac{q_x q_y}{h}\right) + \frac{\partial}{\partial y}\left(\frac{q_y^2}{h} + g\frac{h^2}{2}\right) = -gh\frac{\partial z_b}{\partial y} - g\frac{n^2}{h^{7/3}}|q|q_y \tag{3}$$

where h is the water depth, q_x, q_y and $|q|$ are the two components of the unit discharge and its modulus, z_b is the bed elevation, n is the Manning coefficient, g is the gravity acceleration, R is the rainfall intensity, and i is the infiltration rate.

The hydrodynamic equations are solved with an unstructured finite volume solver, including a specific numerical scheme for hydrological applications, the so-called DHD scheme [26]. The use of this scheme is strongly recommended in rainfall-runoff computations.

2.1.2. SWMM 5.1

The Storm Water Management Model (SWMM) is a 1D dynamic sewer network model developed for the simulation of rainfall-runoff processes and the conveyance of water flows through drainage systems. The sewer network is modelled as a network of links (pipes) connected by nodes (manholes). The hydrodynamics module of SWMM solves the 1D Saint-Venant equations for gradually varied, unsteady flow. These are referred to as dynamic wave analysis and are implemented in the module EXTRAN (Extended Transport) [30].

The SWMM code is split into functions grouped in a single Dynamic-link Library (DLL) [9,24], which simplifies their communication with other software. Each function has a specific task, e.g., start the simulation or advance the simulation step by step.

2.2. Linking Methodology

The structure of SWMM5 allows its interaction and linking with the source code of Iber. The SWMM5-DLL is built from the source code of the OWA-SWMM Open-Source Library [24]. The OWA-SWMM source code is written in C++ and includes a Toolkit API that allows to get and set all the model parameters and hydraulic variables before, during and after the simulation. Once the SWMM-DLL is built, it can be directly invoked from the source code of Iber, and no further actions are needed. Despite new GPU-parallelized releases of Iber are coded in C++ [27], in this work we have used the standard Iber code, which is written in FORTRAN. Therefore, as the OWA-SWMM is written in C++, a standard intrinsic module has been incorporated at the Iber code to stablish the interoperability between the two languages.

The linking methodology includes a set of subroutines that are added to the Iber source code, in order to invoke the SWMM functions and other related operations. In Figure 1, a flow diagram of the 1D/2D linking methodology is presented. Before starting the simulation, each inlet and manhole defined in SWMM must be associated to a surface mesh element in Iber. This can be done automatically (nearest neighbour) or manually by the user. Each inlet and manhole can be assigned to one or more surface elements. This might be appropriate depending on the spatial scale of the study case and the mesh resolution, as it will be shown in the next section.

In a similar way, before the simulation starts, the roof elements defined in Iber are associated to the nearest manhole in SWMM.

Once the simulation starts, at each time step the model solves the surface and sewer flow equations. The surface and sewer network equations are computed by the two solvers in an independent way, using different time steps. The time step in Iber depends on a Courant–Friedrichs–Lewy (CFL) stability constraint that relates the maximum permissible computational time step, the grid size, the flow velocity and the water depth [26]. Similarly, the time step in SWMM is computed by a Courant condition that limited the time step to the time needed for a wave to propagate through the entire pipe [30]. Therefore, a time step synchronization is necessary to guarantee a correct coupling between models and to do the water exchange between minor and major drainage networks at the same elapsed time. The time step of SWMM is usually larger than the Iber time step; thus, the SWMM elapsed time is considered as the synchronization time. The synchronization implies that, at certain computation steps, the Iber time step must be adjusted in order to avoid to exceed the elapsed time of SWMM [9,31]. In order to do so, the computational time step in Iber is defined as follows:

$$\Delta t_{Iber_{n+1}} = \min\left\{T_{syn} + \Delta t_{SWMM_{n+1}} - \sum_{i=1}^{n} \Delta t_{Iber_i}, \Delta t_{Iber\ CFL_{n+1}}\right\} \quad (4)$$

where $\Delta t_{Iber_{n+1}}$ is the time step of Iber for the step $n+1$, adjusted to the synchronization with SWMM if it was necessary, $\Delta t_{SWMM_{n+1}}$ and $\Delta t_{Iber\ CFL_{n+1}}$ are the SWMM and Iber time steps computed independently in both models for the step $n+1$ (i.e., without considering synchronization), T_{syn} is the time of the last synchronization, and $\sum_{i=1}^{n} \Delta t_{Iber_i}$ is the elapsed time of Iber.

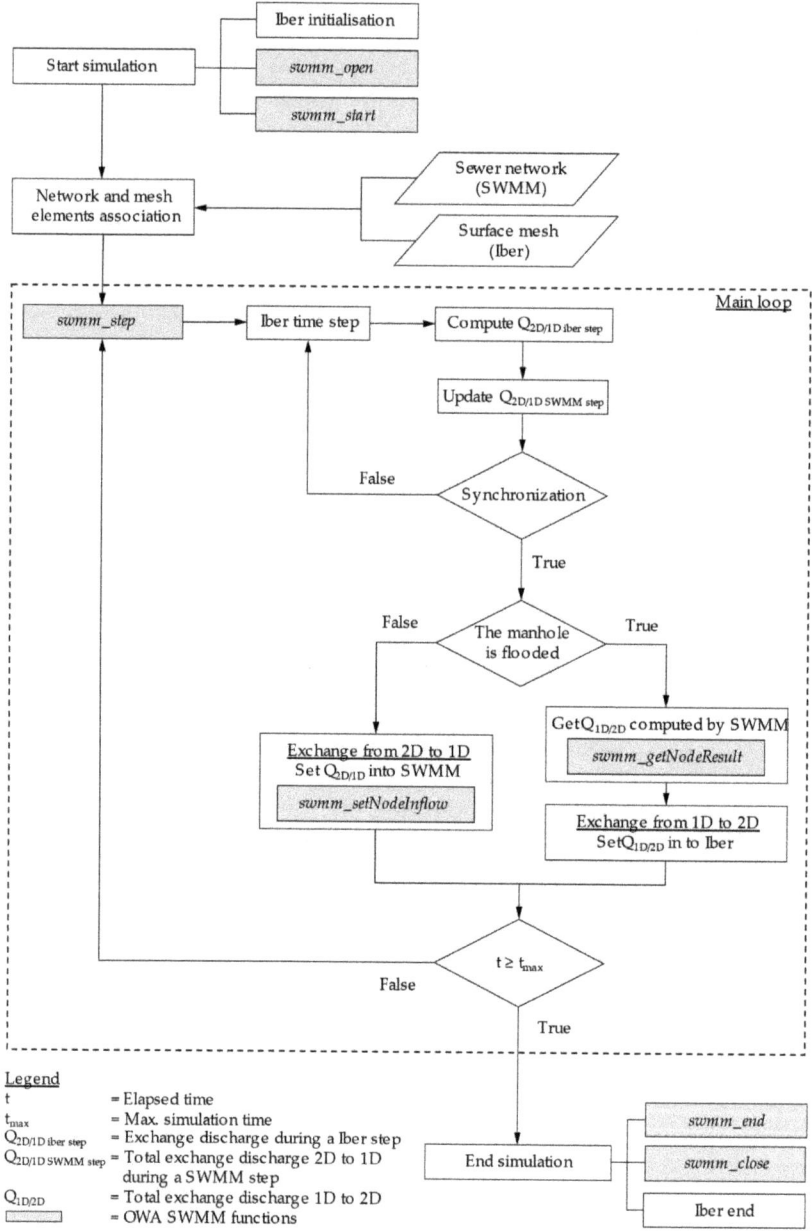

Figure 1. Flow diagram of the 1D/2D urban drainage model.

The water exchange is done every time at which the models are synchronized, assuming a constant water discharge during the whole time step. The interaction between the overland flow and the sewer drainage system occurs only through inlets and manholes. The surface water can only enter the sewer network through the inlets, while the water can only return to the surface through the manholes. This last assumption is reasonable due to when the system is surcharged, the volume of water returned through the inlets is negligible respect the volume returned through the manholes. However, if a manhole is flooded, all the inlets associated to it will not caught any surface flow.

The inlets are directly connected to manholes. In the present version, the conveyance time between the inlet and the manhole is not considered, i.e., the water that enters an inlet is instantaneously directed to the manhole associated to that inlet. This assumption is reasonable since in real applications the conveyance time between inlets and manholes is usually negligible compared to the conveyance time through the major and minor drainage networks. Furthermore, even if detailed studies about the flow in inlets have been performed in the last years [32,33], information about the conduits that link inlets and manholes is usually not available in real case studies.

Roofs are also directly connected to manholes. Thus, all the rain that falls in a roof element is automatically added to the associate manhole regardless the roof height and its conveyance time. Moreover, the roofs will add flow to the manhole node regardless of whether it is flooded or not. It is highlighted that the conveyance time can be included to the model if the information of the roof is available. In addition, the official version will also include the possibility of modelling SuDS (Sustainable Drainage System) techniques such as green roofs or permeable pavements, among others.

To compute the aforementioned interactions between Iber and SWMM, the exchanged discharge introduced in the sewer network is calculated in Iber. Thus, as Iber has a time step smaller than SWMM, for each of the n Iber time steps that are needed to complete a SWMM time step, the cumulative volume of water exchanged is evaluated. Then, when the synchronization occurs, the functions of SWMM are invoked to introduce the volume caught by the inlets in the sewer between the previous and the current synchronization time as a manhole discharge. Free weir, submerged weir and orifice equations used in previous studies [9,10,31] have been implemented to compute the flow interchange depending on the water level at the surface mesh element, the mesh element elevation and the hydraulic head at the manhole. On the other hand, if a manhole is flooding, the node flood discharge is computed by SWMM and transferred to the surface as an Iber source. It is highlighted that the same water volume that is extracted from Iber is introduced in SWMM and, vice versa. Finally, SWMM step is computed establishing the next synchronization time and it will not run again until Iber reaches this synchronization time and new exchange occurs.

3. Case Studies and Results

In order to validate the dual drainage model presented in Section 2 and to show its capabilities, three case studies have been modelled. The case studies were chosen in order to assess numerical aspects such as mass conservation and numerical stability, to validate model output against laboratory experimental data and to show the workflow methodology to set up a model in a real application. The study cases include free surface and surcharged flow conditions in the drainage network as well as water discharge exchanges in both directions, i.e., from major to minor network and vice versa. All the tests have been solved using the DHD scheme of Iber and a wet-dry threshold of 0.1 mm.

3.1. Simplified Urban Street

3.1.1. Case Study Description

This case study consists in a synthetic urban street, and it is aimed to verify some numerical aspects and basic capabilities of the model. The spatial domain includes a roadway with pavements at both sides, a pedestrian area, a green area and buildings (Figure 2a). The roadway is 40 m in length and 7 m in width, and it is separated from the 2-m wide pavements by a 15-cm high curb. The roadway and pavements have 2% and 1% slopes in the transversal direction, respectively, and 1% longitudinal slope. The buildings are 10 m in width and are directly connected to the manholes. The sewer system and the surface are connected by 8 inlets and 4 manholes. Each inlet is connected to its nearest manhole. Manholes M1, M2 and M3 are connected by a pipe with an inner diameter of 500 mm, while the pipe that connects manholes M4 and M2 has in inner diameter of 300 mm. Both pipes have a slope of 1%.

The network outlet (O1) is located 10 m downstream from manhole M3, following the pipe slope, and its invert elevation is −2 m (Figure 3). The invert manhole relative depth regarding the surface is −2 m for M1, M2 and M3 and −2.05 m for M4. At O1, the reference elevation for the head is the surface elevation, 0 m in this location.

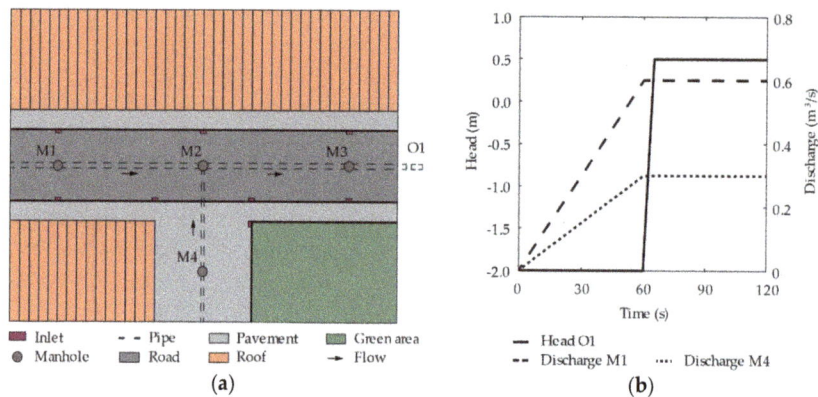

Figure 2. (a) Scheme of the urban street; (b) boundary conditions at manholes and outfall.

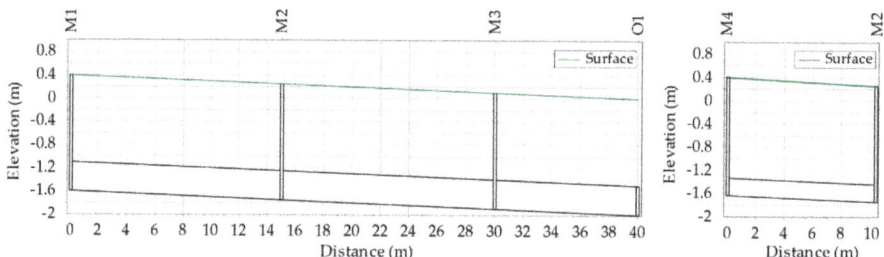

Figure 3. Longitudinal profile of the drainage network.

In order to simulate surcharged flow and surface flood conditions, two inlet hydrographs were imposed at manholes M1 and M4, and a hydraulic head condition was fixed at the outlet O1, as shown in Figure 2b. A constant rainfall intensity of 80 mm/h was imposed in all the spatial domain. The surface domain was discretized with an unstructured mesh with an average element size of 0.3 m and approximately 32,000 elements. The Manning coefficient was set to 0.016 for the pavement and road surfaces, which are both considered impervious (no infiltration), and 0.032 for the green area. In the green area, a constant infiltration rate of 10 mm/h was considered after an initial abstraction of 10 mm.

3.1.2. Results

Figure 4 shows the evolution of water depths and discharges at the four manholes. At all the manholes, the water depth increases slowly during the first 60 min, due to the rainfall input and to the water coming from the imposed hydrographs at manholes M1 and M4. At that time, the sudden rise of the downstream boundary hydraulic head condition triggers a sharp increase in the manhole depths. Once their maximum depth is exceeded, the water flows outside the manhole and into the street surface. Figure 5 shows the computed maximum water depths in the street surface. It can be seen the water column that typically occurs when flooding from manholes.

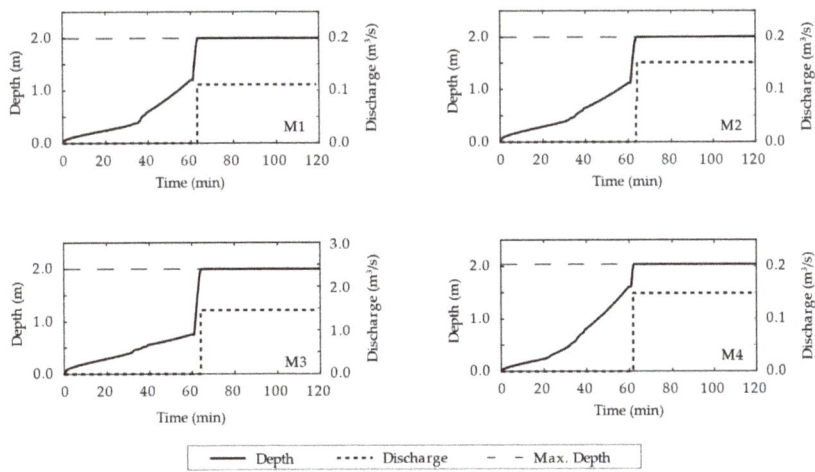

Figure 4. Results of depths and discharges at manholes and maximum water depth for manholes.

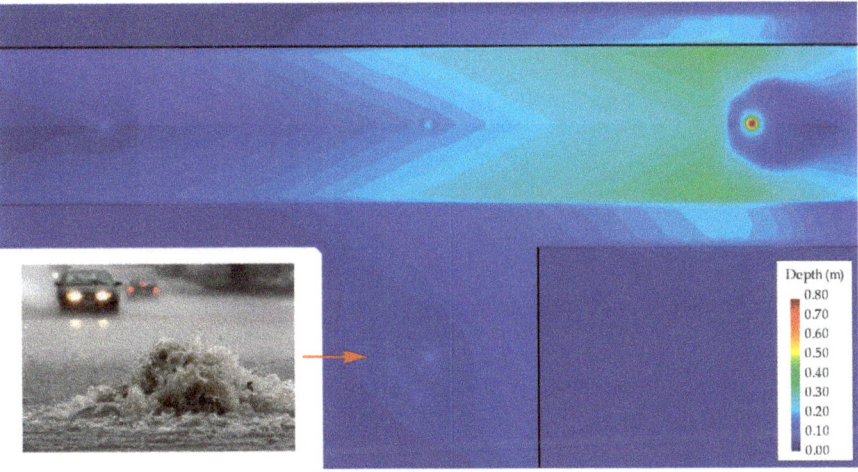

Figure 5. Results of maximum depths at surface and photography of a manhole flooding [34] that justifies the assumption that the main returned flow to the surface occurs through the manholes.

The effect of rainfall alone is not enough to trigger the surface flooding in this case. However, the rainfall runoff allows to check for this case study the behaviour of the inlets before the flooding occurs. Figure 6a shows the volume of water intercepted by inlets and roofs. It can be noticed how, once the flooding occurs (approximately at minute 60), the inlets stop catching water.

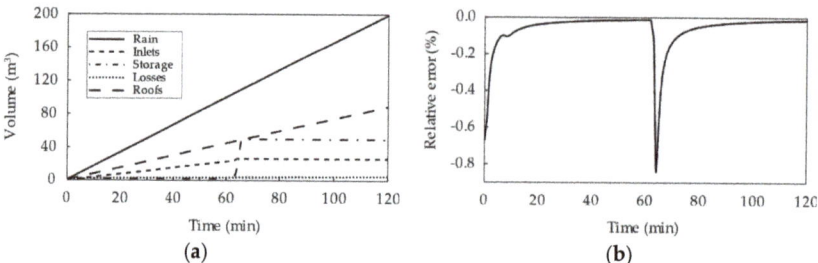

Figure 6. (a) Time series evolution of accumulative volumes; (b) mass balance error during the simulation for the simplified urban street.

In order to verify the global mass conservation of the coupled model, including the surface and sewer models as well as the linking methodology, the mass balance error was computed at each time step of the simulation as

$$error = (P - S - V_{inlet} - V_{roofs} - V_{boundary} - V_{infiltration} + V_{flood}) \qquad (5)$$

where P is the precipitation volume, S is the overland flow storage, V_{inlet} is the volume intercepted by the inlets, V_{roofs} is the volume intercepted by the roofs, $V_{boundary}$ is the outlet volume through the domain boundaries, $V_{infiltration}$ is the infiltration volume, and V_{flood} is the volume that returns to the surface through the manholes. The relative accumulated mass balance error in relation with the outlet boundary volume is at every time step lower than 1% (Figure 6b).

3.2. Full-Scale Urban Drainage Physical Model

3.2.1. Case Study Description

This case presents the experimental validation of the dual drainage model using a set of laboratory data obtained in a real scale physical model. The physical model consists of a full-scale street section of 36 m² with a sewer system and a rainfall simulator. This facility was used in previous studies [10,35–37] to validate urban drainage models and to measure wash-off and sediment transport in urban environments. The street consists of a concrete roadway and a concrete pavement separated by a 15-cm high curb. The roadway has 2% and 0.5% transversal and longitudinal slopes, respectively. Surface runoff is drained through two inlets and through a lateral channel that ends into a third inlet. From there, water is conveyed through the sewer system to a downstream outfall (Figure 7). The street geometry and experimental data presented in WASHTREET [38] were used to build the numerical model.

Three different rainfall intensities of 30, 50 and 80 mm/h were simulated. In order to compare the experimental data with the numerical results, the hydraulic variables in a set of control points were analysed (Figure 7b). The water discharges through the two inner inlets and at the outfall were compared. In addition, the water depth at 6 locations within the pipes and at 3 surface control points were compared.

The surface domain was discretized using an unstructured mesh with an average element size of 0.06 m and approximately 20,000 elements. From previous studies in this laboratory facility [36], the Manning coefficient was set to 0.016, and an initial abstraction of 0.6 mm was established in the whole surface. In the lateral plastic channel and in all the pipes of the sewer network, the Manning coefficient was set to 0.008 [10].

Figure 7. (**a**) Physical model facility and (**b**) measuring points in inlets, pipes and surfaces used for results validation.

3.2.2. Results

Figure 8 shows the experimental and numerical hydrographs computed at the two inlets and at the outfall for the three rainfall intensities. The numerical and experimental data present a good agreement at both inlets being the mean absolute error (MAE) less than 0.01 L/s for all the rainfall intensities. Regarding the agreement at the outfall, a small-time lag between the numerical and experimental data can be observed during the rising limb of the hydrographs (MAE less than 0.09 L/s). This difference is due to the way in which the outfall discharge was measured in the experimental tests. Experimental discharge was measured in a deposit located at the end of the sewer network that produced a slight lamination of the experimental hydrograph [37].

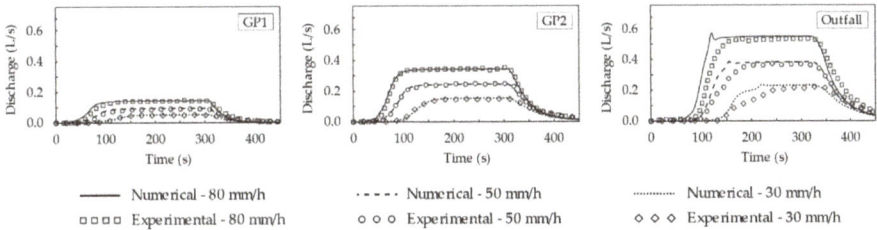

Figure 8. Experimental and numerical profiles of discharges at inlets GP1 (**left**) and GP2 (**middle**) and at the outfall (**right**).

Figure 9 shows the water depths at pipes and at the surface. The results present a reasonable agreement at all the control points, specially at P1, P2 and P4, being the MAE less than 1 mm for all the rainfall intensities except for P2 and intensity 30 mm/h that is 2 mm. P3 present less agreement being the maximum MAE 3.5 mm. Notice that the water depth at those control points is very low in relation to the diameter of the pipe, so a minimum difference between the real and modelled geometry can significantly affect the results. The same happens at the surface control points, where the measured water depths are extremely small, and thus, the effect of the microtopography and other physical phenomena that are no considerer in the numerical model, such as drop impacts, can have an effect on the numerical-experimental agreement. Nonetheless, the numerical model simulates the surface water depths and its temporal dynamics with a reasonable agreement (MAE always less than 1.5 mm).

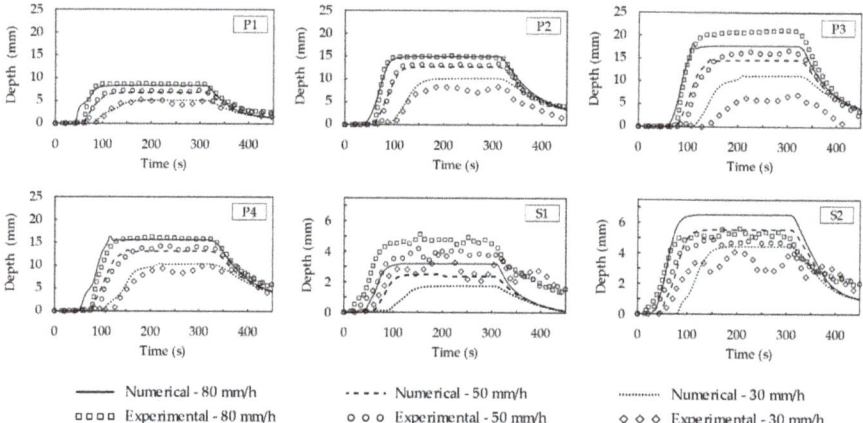

Figure 9. Experimental and numerical profiles of depths in pipes at control points P1 (**top-left**), P2 (**top-middle**), P3 (**top-right**) and P4 (**bottom-left**) and in surface at control points S1 (**bottom-middle**) and S2 (**bottom-right**).

Finally, Figure 10a shows the water volumes computed during the simulation. Most of the rainfall volume is drained through the inlets. Using the mass conservation Equation (5), the mean mass conservation error during the whole simulation in relation with the precipitation volume is 0.003% and tends to zero at the end of simulation (Figure 10b).

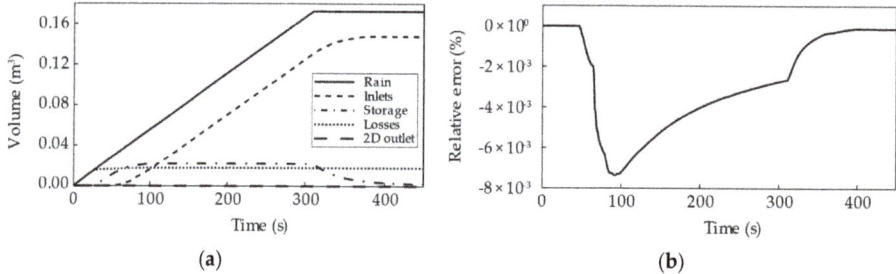

Figure 10. (**a**) Time series evolution of accumulative volumes; (**b**) mass balance error during the simulation for full-scale urban drainage physical model.

3.3. Real Urban Settlement

3.3.1. Case Study Description

The last case study shows the application of the dual drainage model to El Rubio, an urban settlement located in Andalucía (Spain) with a population of 3500 inhabitants. The workflow needed to import the sewer network from a GIS data base using external free tools is also detailed.

The urban area was split in two different regions according to the different land uses: an urban impervious region and a pervious region. The available digital terrain model (DTM), which has a resolution of 2 m, and other geometric information such as the layout of buildings and infiltration areas are shown in Figure 11. Due to the lack of field data that could be used to calibrate the model parameters, the Manning coefficient was set to 0.016 in the urban area and to 0.20 in the rest of the domain [39,40]. The SCS Curve Number Method was used to compute infiltration losses. Based on the GCN250 dataset [41], a curve-number (CN) of 75 was imposed for all the pervious areas, while the urban domain was considered to be completely impervious. The spatial domain included in the computations includes all the surrounding terrains that drain into the urban region, in order to consider the surface runoff coming from upstream areas.

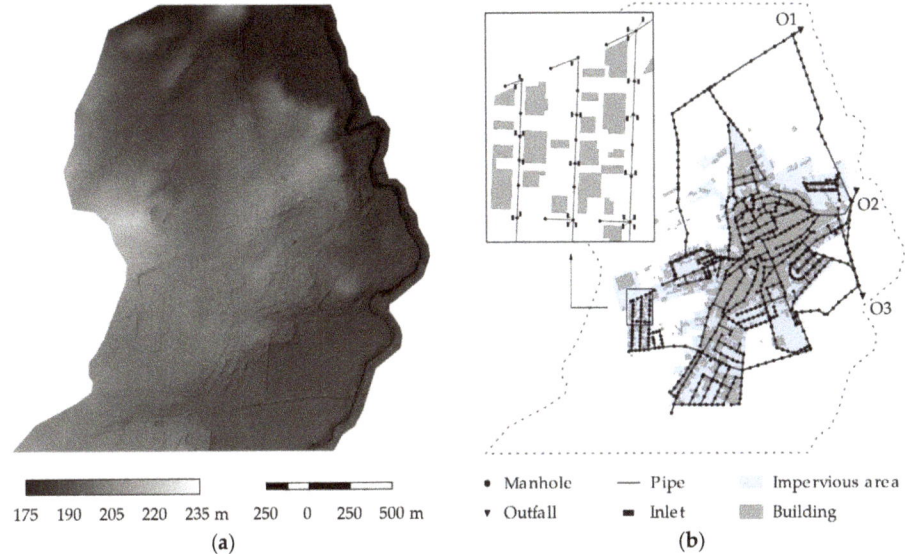

Figure 11. (a) DEM of the study area; (b) sewer network of the urban settlement.

The sewer system (Figure 11b) is composed of 727 manholes, 753 links and 3 outfalls. The information of the sewer system was provided in GIS data format, so it was necessary to convert it to the SWMM input format. To do so, the R software package *swmmr* was used. The R package *swmmr* is a free tool that contains specific functions to read and write SWMM files, run SWMM simulations from R and convert SWMM input files to and from GIS data [42]. The inlets were not included in the SWMM input file. Thus, all the inlets that do not have an associated manhole in the GIS data are automatically assigned by the model to the nearest manhole. Unlike in the two previous case studies, the DTM available for this case has a limited resolution (2 m), which means that the pathlines of water at the streets could not be resolved with an enough resolution to direct the water towards the inlets. Other methodologies to catch overland flow when the topography does not have enough resolution, such as the definition of a density of inlets in a micro-catchment that drains into a manhole [33], will be analysed in future works.

In order to evaluate the effect of the numerical discretisation on the results, two different computational meshes with different spatial resolution were used to discretize the domain. The coarse mesh has 100,000 elements and an average element size of 10 m, while the fine mesh has 1,000,000 elements, with an average element size of 2 m inside the urban area and of 20 m in the rest of the domain.

A synthetic hyetograph with a duration of 1 h and a return period of 25 years was computed from regional intensity–duration–frequency curves (IDF) [43] using the alternating block method (Figure 12b). The hyetograph was imposed homogeneously in space.

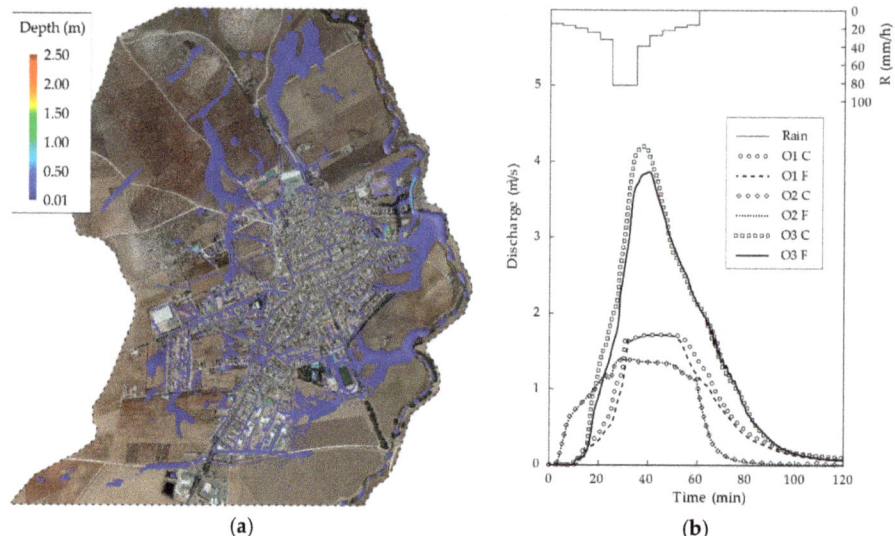

Figure 12. (a) Maximum depths at surface; (b) numerical hydrographs computed at outfalls for a 1 h design storm with a 25-year return period for coarse (C) and fine (F) mesh.

3.3.2. Results

Figure 12a shows the spatial distribution of the maximum water depth computed during the simulation with the fine mesh. To facilitate the visualization of the results, water depths lower than 0.01 m are not represented. The results obtained with the coarse mesh are very similar and are therefore not shown in the figure.

Figure 12b shows the discharges at the three outfalls (see location in Figure 11b) computed with the fine and coarse meshes. The hydrographs computed at outfall O2 with both meshes are virtually the same (MAE of 0.01 m^3/s). At outfall O1, there are some minor differences at some time steps, but those are not significant for practical purposes (MAE of 0.06 m^3/s), and the peak discharge computed with both meshes is the same. Regarding outfall O3, the MAE increases to 0.11 m^3/s, and the difference in peak discharge is around 8%. While these differences are larger than in the case of O1 and O2, they are still small for the analysis of sewer networks in real applications, where the uncertainties on input data are in general larger than those values. Moreover, it should be considered that the CPU time was 15 times larger when using the fine mesh in relation to the coarse mesh.

A simulation without the sewer network was carried out in order to highlight the role of the sewer drainage system during a rainfall event. Figure 13 shows the maps of maximum depth during the rainfall event considering and not considering the sewer network in the model. The results are shown only in the urban impervious area, and the building shapes were removed from the figure in order to facilitate the visualization. The extension of the flood is significantly different, being, as expected, bigger for the case without sewer network, which corroborates the essential role that play the sewer network systems during extreme rainfall events. Thus, the no consideration of the sewer network implies a 10% and 35% of increase in the depths and in the flood extension, respectively.

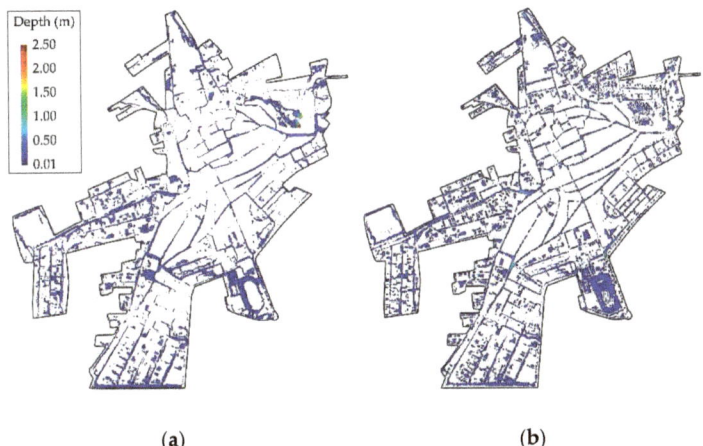

Figure 13. Maximum depths on the surface: (**a**) with and (**b**) without considering the sewer network.

4. Conclusions

A 1D/2D dual drainage model linking two free distributed hydraulic models (Iber and SWMM) was presented. The model allows the simulation of water flow in the surface and sewer drainage networks, including their bidirectional interaction. The linking methodology computes the discharge transferred from the surface to the sewer network, and vice versa, at each computational time step and is based in a SWMM-DLL library that contains different functions to interact with SWMM before, during and after the simulation.

The capabilities of the model were shown in three case studies at different spatial scales. The first case study was used to validate the basic capabilities of the model and to show its suitability to compute surface flooding conditions caused by the surcharge of the sewer network. In the second case study, the numerical results are compared against experimental data obtained in a full-scale urban drainage physical model. Finally, the third case was used to show the capability of the model to simulate pluvial floods in real urban settlements, incorporating the available GIS data of the sewer system.

Further developments of the model will include different methodologies to compute the transfers of water captured by inlets and roofs, the implementation of SuDS (Sustainable Drainage System) techniques and the enhancement of the computational time. In addition, future works will also compare the present model with other well-known dual drainage models, in order to analyse its capabilities and its advantages and disadvantages against these.

Author Contributions: Conceptualization, L.C., J.P. and E.S.; methodology, L.C., J.P. and E.S.; software development, E.S.; supervision software development, L.C. and J.P.; validation, E.S.; writing—original draft preparation, E.S.; writing—review and editing, L.C. and J.P.; supervision, L.C. and J.P. All authors have read and agreed to the published version of the manuscript.

Funding: The contract of the first author is funded by the INTERREG ATLANTIC AREA program through the project AA-FLOODS Enhanced Prevention, Warning, Coordination and Emergency Management Tools for Floods at Local Scales (EAPA_45/2018).

Acknowledgments: The authors acknowledge ARECIAR and CMAOT for supplying, respectively, the sewer network information and the topographic data of the third test case.

Conflicts of Interest: The authors declare no conflict of interest. The funders had no role in the design of the study; in the collection, analyses or interpretation of data; in the writing of the manuscript or in the decision to publish the results.

References

1. Cea, L.; Garrido, M.; Puertas, J.; Jácome, A.; Del Río, H.; Suárez, J. Overland flow computations in urban and industrial catchments from direct precipitation data using a two-dimensional shallow water model. *Water Sci. Technol.* **2010**, *62*, 1998–2008. [CrossRef]
2. Fraga, I.; Cea, L.; Puertas, J.; Suarez, J.; Jimenez, V.; Jacome, A. Global sensitivity and GLUE-Based uncertainty analysis of a 2D-1D dual urban drainage model. *J. Hydrol. Eng.* **2016**, *21*. [CrossRef]
3. Chen, A.S.; Leandro, J.; Djordjević, S. Modelling sewer discharge via displacement of manhole covers during flood events using 1D/2D SIPSON/P-DWave dual drainage simulations. *Urban Water J.* **2016**, *13*, 830–840. [CrossRef]
4. Martins, R.; Leandro, J.; Chen, A.S.; Djordjević, S. A comparison of three dual drainage models: Shallow water vs local inertial vs diffusive wave. *J. Hydroinform.* **2017**, *19*, 331–348. [CrossRef]
5. Djordjević, S.; Prodanović, D.; Maksimović, Č. An approach to simulation of dual drainage. *Water Sci. Technol.* **1999**, *39*, 95–103. [CrossRef]
6. Leandro, J.; Chen, A.S.; Djordjevi, S.; Savi, D.A. Comparison of 1D/1D and 1D/2D coupled (Sewer/Surface) hydraulic models for urban flood simulation. *J. Hydraul. Eng.* **2009**, *135*, 495–504. [CrossRef]
7. Fraga Cadórniga, I. Desarrollo de un modelo dual 1D/2D para el cálculo del drenaje urbano: Modelo numérico y validación experimental. Ph.D. Thesis, Universidade da Coruña, A Coruña, Spain, 2015.
8. Aragón-Hernández, J.L. Modelación numérica integrada de los procesos hidráulicos en el drenaje urbano. Ph.D. Thesis, Universitat Politècnica de Catalunya, Barcelona, Spain, 2013.
9. Leandro, J.; Martins, R. A methodology for linking 2D overland flow models with the sewer network model SWMM 5.1 based on dynamic link libraries. *Water Sci. Technol.* **2016**, *73*, 3017–3026. [CrossRef]
10. Fraga, I.; Cea, L.; Puertas, J. Validation of a 1D-2D dual drainage model under unsteady part-full and surcharged sewer conditions. *Urban Water J.* **2017**, *14*, 74–84. [CrossRef]
11. Rossman, L.A. *Storm Water Management Model User's Manual Version 5.1.*; EPA/600/R-14/413b; National Risk Management Research Laboratory, Office of Research and Development, U.S. Environmental Protection Agency: Cincinnati, OH, USA, 2015; p. 352.
12. Hsu, M.H.; Chen, S.H.; Chang, T.J. Dynamic inundation simulation of storm water interaction between sewer system and overland flows. *J. Chin. Inst. Eng. Trans. Chin. Inst. Eng. Ser. A Chung Kuo K. Ch'eng Hsuch K'an* **2002**, *25*, 171–177. [CrossRef]
13. Pathirana, A.; Tsegaye, S.; Gersonius, B.; Vairavamoorthy, K. A simple 2-D inundation model for incorporating flood damage in urban drainage planning. *Hydrol. Earth Syst. Sci.* **2011**, *15*, 2747–2761. [CrossRef]
14. Jahanbazi, M.; Egger, U. Application and comparison of two different dual drainage models to assess urban flooding. *Urban Water J.* **2014**, *11*, 584–595. [CrossRef]
15. Kim, S.E.; Lee, S.; Kim, D.; Song, C.G. Stormwater inundation analysis in small and medium cities for the climate change using EPA-SWMM and HDM-2D. *J. Coast. Res.* **2018**, *85*, 991–995. [CrossRef]
16. Martins, R.; Leandro, J.; Djordjević, S. Influence of sewer network models on urban flood damage assessment based on coupled 1D/2D models. *J. Flood Risk Manag.* **2018**, *11*, S717–S728. [CrossRef]
17. Adeogun, A.G.; Pathirana, A.; Daramola, M.O. Others 1D-2D hydrodynamic model coupling for inundation analysis of sewer overflow. *JEAS J. Eng. Appl. Sci.* **2012**, *7*, 356–362. [CrossRef]
18. Delelegn, S.W.; Pathirana, A.; Gersonius, B.; Adeogun, A.G.; Vairavamoorthy, K. Multi-objective optimisation of cost-benefit of urban flood management using a 1D2D coupled model. *Water Sci. Technol.* **2011**, *63*, 1053–1059. [CrossRef] [PubMed]
19. Seyoum, S.D.; Vojinovic, Z.; Price, R.K.; Weesakul, S. Coupled 1D and noninertia 2D flood inundation model for simulation of urban flooding. *J. Hydraul. Eng.* **2012**, *138*, 23–34. [CrossRef]
20. Phillips, B.C.; Yu, S.; Thompson, G.R.; Silva, N. De 1D and 2D Modelling of Urban Drainage Systems using XP-SWMM and TUFLOW. In Proceedings of the 10th International Conference on Urban Drainage, Copenhagen, Denmark, 21–26 August 2005.
21. Bisht, D.S.; Chatterjee, C.; Kalakoti, S.; Upadhyay, P.; Sahoo, M.; Panda, A. Modeling urban floods and drainage using SWMM and MIKE URBAN: A case study. *Nat. Hazards* **2016**, *84*, 749–776. [CrossRef]
22. SWMM5 Modeling with PCSWMM. Available online: https://www.pcswmm.com/ (accessed on 6 September 2020).

23. Bladé, E.; Cea, L.; Corestein, G.; Escolano, E.; Puertas, J.; Vázquez-Cendón, E.; Dolz, J.; Coll, A. Iber: Herramienta de simulación numérica del flujo en ríos. *Rev. Int. Metod. Numer. Calc. Disen. Ing.* **2014**, *30*, 1–10. [CrossRef]
24. SWMM-Docs: Open Water Analytics Stormwater Management Model. Available online: http://wateranalytics.org/Stormwater-Management-Model/index.html (accessed on 17 July 2020).
25. Bladé Castellet, E.; Cea, L.; Corestein, G. Numerical modelling of river inundations. *Ing. Agua* **2014**, *18*, 68. [CrossRef]
26. Cea, L.; Bladé, E. A simple and efficient unstructured finite volume scheme for solving the shallow water equations in overland flow applications. *J. Am. Water Resour. Assoc.* **2015**, *5*, 2. [CrossRef]
27. García-Feal, O.; González-Cao, J.; Gómez-Gesteira, M.; Cea, L.; Domínguez, J.M.; Formella, A. An accelerated tool for flood modelling based on Iber. *Water* **2018**, *10*, 1459. [CrossRef]
28. Fraga, I.; Cea, L.; Puertas, J. Effect of rainfall uncertainty on the performance of physically based rainfall–runoff models. *Hydrol. Process.* **2019**, *33*, 160–173. [CrossRef]
29. Cea, L.; Legout, C.; Darboux, F.; Esteves, M.; Nord, G. Experimental validation of a 2D overland flow model using high resolution water depth and velocity data. *J. Hydrol.* **2014**, *513*, 142–153. [CrossRef]
30. Rossman, L.A. *Storm Water Management Model Reference Manual Volume II—Hydraulics*; U.S. Environmental Protection Agency: Washington, DC, USA, 2017.
31. Chen, A.S.; Djordjević, S.; Leandro, J.; Savić, D.A. The urban inundation model with bidirectional flow interaction between 2D overland surface and 1D sewer networks. *Novatech 2007* **2007**, 465–472.
32. Lopes, P.; Leandro, J.; Carvalho, R.F.; Páscoa, P.; Martins, R. Numerical and experimental investigation of a gully under surcharge conditions. *Urban Water J.* **2015**, *12*, 468–476. [CrossRef]
33. Gómez, M.; Russo, B. Methodology to estimate hydraulic efficiency of drain inlets. *Proc. Inst. Civ. Eng. Water Manag.* **2011**, *164*, 81–90. [CrossRef]
34. Flood risk management. Engineering. University of Exeter. Available online: https://emps.exeter.ac.uk/engineering/research/cws/research/flood-risk/rapids.html (accessed on 30 July 2020).
35. Naves, J. Wash-off and Sediment Transport Experiments in a Full-Scale Urban Drainage Physical Model. Ph.D. Thesis, Universidade da Coruña, A Coruña, Spain, 2019.
36. Naves, J.; Anta, J.; Puertas, J.; Regueiro-Picallo, M.; Suárez, J. Using a 2D shallow water model to assess Large-Scale Particle Image Velocimetry (LSPIV) and Structure from Motion (SfM) techniques in a street-scale urban drainage physical model. *J. Hydrol.* **2019**, *575*, 54–65. [CrossRef]
37. Naves, J.; Anta, J.; Suárez, J.; Puertas, J. Hydraulic, wash-off and sediment transport experiments in a full-scale urban drainage physical model. *Sci. Data* **2020**, *7*, 1–13. [CrossRef]
38. Naves, J.; Anta, J.; Suárez, J.; Puertas, J. Washtreet—Hydraulic, wash-off and sediment transport experimental data obtained in an urban drainage physical model. *Sci. Data* **2019**. [CrossRef]
39. Cea, L.; Legout, C.; Grangeon, T.; Nord, G. Impact of model simplifications on soil erosion predictions: Application of the GLUE methodology to a distributed event-based model at the hillslope scale. *Hydrol. Process.* **2016**, *30*, 1096–1113. [CrossRef]
40. Fraga, I.; Cea, L.; Puertas, J. Experimental study of the water depth and rainfall intensity effects on the bed roughness coefficient used in distributed urban drainage models. *J. Hydrol.* **2013**, *505*, 266–275. [CrossRef]
41. Jaafar, H.H.; Ahmad, F.A.; Beyrouthy, N. El GCN250, new global gridded curve numbers for hydrologic modeling and design. *Sci. Data* **2019**, 1–9. [CrossRef]
42. Leutnant, D.; Döring, A.; Uhl, M. Swmmr—An R package to interface SWMM. *Urban Water J.* **2019**, *16*, 68–76. [CrossRef]
43. Arias, J.S. *Máximas Lluvias Diarias en España Peninsular*; Seria Monográfica del Ministerio de Fomento: Madrid, Spain, 1999; p. 55.

© 2020 by the authors. Licensee MDPI, Basel, Switzerland. This article is an open access article distributed under the terms and conditions of the Creative Commons Attribution (CC BY) license (http://creativecommons.org/licenses/by/4.0/).

Article

Multistep Flood Inundation Forecasts with Resilient Backpropagation Neural Networks: Kulmbach Case Study

Qing Lin *, Jorge Leandro, Stefan Gerber and Markus Disse

Chair of Hydrology and River Basin Management, Department of Civil, Geo and Environmental Engineering, Technical University of Munich, Arcisstrasse 21, 80333 Munich, Germany; jorge.leandro@tum.de (J.L.); stefan.gerber@tum.de (S.G.); markus.disse@tum.de (M.D.)
* Correspondence: tsching.lin@tum.de; Tel.: +49-89-289-23228

Received: 21 October 2020; Accepted: 10 December 2020; Published: 19 December 2020

Abstract: Flooding, a significant natural disaster, attracts worldwide attention because of its high impact on communities and individuals and increasing trend due to climate change. A flood forecast system can minimize the impacts by predicting the flood hazard before it occurs. Artificial neural networks (ANN) could efficiently process large amounts of data and find relations that enable faster flood predictions. The aim of this study is to perform multistep forecasts for 1–5 h after the flooding event has been triggered by a forecast threshold value. In this work, an ANN developed for the real-time forecast of flood inundation with a high spatial resolution (4 m × 4 m) is extended to allow for multiple forecasts. After trained with 120 synthetic flood events, the ANN was first tested with 60 synthetic events for verifying the forecast performance for 3 h, 6 h, 9 h and 12 h lead time. The model produces good results, as shown by more than 81% of all grids having an RMSE below 0.3 m. The ANN is then applied to the three historical flood events to test the multistep inundation forecast. For the historical flood events, the results show that the ANN outputs have a good forecast accuracy of the water depths for (at least) the 3 h forecast with over 70% accuracy (RMSE within 0.3 m), and a moderate accuracy for the subsequent forecasts with (at least) 60% accuracy.

Keywords: hazard; artificial neural network; resilient backpropagation; multistep urban flood forecast

1. Introduction

Floods are one of the most threatening hazards to civilian safety and infrastructures, causing damages and losses over the world [1]. Especially in densely populated urban areas, urbanization, aging of drainage systems and climate change contribute to growing flood risk in many countries. Ione important mitigation measure is the prediction of future flood occurrences. Real-time flood forecast with sufficient lead time can boost the use of preventive measures for flood mitigation. Such measures can minimize the threats to communities and individuals at risk of flooding [2].

Various types of hydrology and hydraulic models are available for flood forecasting [3]. From the different modeling types, these can be classified into rainfall-runoff models, one-dimensional (1D) model, two-dimensional (2D) model, and coupled 1D–1D model and 1D–2D model. From the forecast application perspective, 2D and 1D–2D models are able to provide directly a spatial surface flood representation, which is essential for flood damage estimation. 1D–1D models, on the other hand, relies heavily on GIS pre- and post-treatments. Most 2D hydrodynamic models are computationally expensive [4]. Even with the help of up-to-date computational techniques for 2D simulations, the computational capability is still inadequate for a real-time forecast [5].

Data-driven is a fast-growing alternative to hydrodynamic models due to the development of computing technologies in recent years. Data-driven models ignore the physical background of a problem and rather explore the relation between the input and output data [6]. For short or long-term flood forecasts, different data-driven models have been implemented, such as neuro-fuzzy [7], support vector machine [8], support vector regression [9,10], Bayesian linear regression methods [11] and artificial neural network (ANN) [12]. Among them, artificial neural networks (ANN) can be an effective tool for flood modeling, if it is properly applied, overcoming pitfalls as over-fitting/under-fitting with sufficient and representative data for model training [13]. Dawson and Wilby applied ANN to conventional hydrological models in flood-inclined catchments in the UK in 1998 [14]. After that, numerous examinations about flood forecasts in catchment scales emerged [2], [15]. Sit and Demir [16] integrated the river network spatial information to improve the forecast accuracy in Iowa. Bustami et al. applied the backpropagation ANN model for forecasting water levels at gauging stations [17]. ANN showed its great potential on the short-time forecast of extreme water levels with a comparable accuracy to the physical model with a far less computational cost [18]. Simon Berkhahn et al. applied an ANN for two-dimensional (2D) urban pluvial inundation extent forecast [19]. Lin et al. applied backpropagation networks for maximum flood inundation extent prediction and achieved a comparable accuracy to the hydraulic model [20].

The objective of this work is to develop a multistep flood forecast method for urban areas at a fine spatial resolution of 4 m by 4 m. Unlike Lin et al. [20], in this study, an ANN-based framework is proposed for performing multistep forecasts for 1–5 h. To the author's knowledge, only the works of Chang et al. [2] and Shen and Chang [21] were able to produce multistep flood forecasting maps. The novelty herein is the forecast at a finer resolution of 4 m × 4 m, suitable for urban flood forecast. Our hypothesis is that the ANN forecast model can provide comparably accurate flood extents as hydraulic models, but with a much shorter time of several seconds. As the hydrodynamics-based model runs usually takes several hours, the reduction in forecast time of ANN enables more flood mitigation measures to become effective. In Section 2, we introduce the resilient backpropagation artificial neural network structure and the validation of the model. Section 3 describes the study area and the data preparation of the event database. Section 4 shows the results of flood forecasts of the first intervals for synthetic and historical flood events and the real-time forecast for historical flood events. Sections 5 and 6 are the discussion and conclusion of this work.

2. Methods

2.1. Data and Structure of Artificial Neural Networks

Artificial neural networks are algorithms applied to map features into a series of outputs. Through a structure of the input, output and intermediate hidden layers, artificial neural networks can learn data relationships between input and output data [22]. A feedforward neural network is applied in this work for modeling the study area, proceeding and transmitting data in a network structure [23]. One of the most widely used ANN is the multilayer perceptron (MLP) [24]. The MLP consists of highly interconnected neurons organized in layers to process information. The neurons in one layer are fully connected to each neuron in the next layer. Each connection is then assigned a weight. Each neuron collects values from the previous layer by summing up the results from the previous neuron values multiplying the weight on each input arcs and storing the results on itself. An activation function is used to transfer the results from the hidden layers to the output layer, and a loss function is applied to measure the fit of the neural network to a set of input–output data pair.

In this case study, the input layer collects the seven inflows to the urban area of Kulmbach, given the hourly discharge intensity. The output layer is fed with the hourly raster inundation map with a resolution of 4 m × 4 m from the event database. Between the input layer and the output layer, the ANN has 2 hidden layers with 10 nodes per layer. One hundred twenty synthetic events from the event database are used for network training (see Section 3.2). Afterward, the other 60 events in the event database are used for model validation. This represents 2/3 of the date for training and 1/3 for testing, which is often found in the literature [25]. Finally, the model is applied to forecast three historical events. The widely applied sigmoid function is chosen as the activation function for the neural network [26]. Due to the high-resolution of the map (4 m by 4 m), weights between the last hidden layer and the output layer would have been 1 GB RAM with a dimension of the problem of more than 30 million. The optimization of these weights is very time-consuming, even with the latest optimization techniques [27]. Hence, a "divide and conquer" strategy is used to enable calculation in a single PC. In addition, it should be noted that, in principle, the results of the two strategies should be the same. Some alternative artificial network structures, such as a convolutional neural network (CNN), could not be applied in this study. The network size would require a very large number of hypermeters, which was beyond the memory capacity of a personal computer for forecasting purposes [28]. Furthermore, the time for training can be reduced with our strategy due to parallelization. The estimated time for training all networks in parallel with four cores is 6 h. To reduce the training time and save memory requirements, the study area is divided into 50 × 50 squared grids (see Figure 1). A similar idea of splitting has been applied to a former study [19]. Each grid had four independent ANNs for intervals (3 h, 6 h, 9 h and 12 h). In total, 10,000 ANNs are trained to produce multistep forecasts.

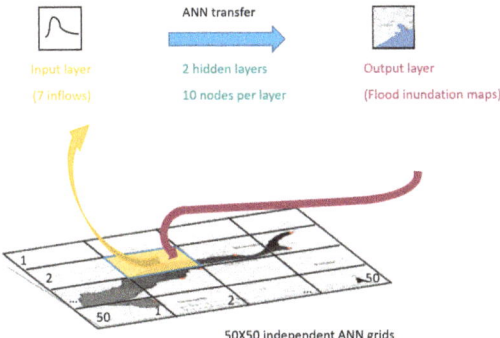

Figure 1. The forward-feed neural network setup in the forecast study. The input layer is fed with discharge inflows of certain time interval windows. The output layer generates the flood inundation for that interval. Resilient backpropagation is applied for training this network.

2.2. Hyperparameter Tuning in ANN

To optimize weights in ANN, resilient backpropagation is a widely applied effective algorithm [29]. According to Shamim et al. and Panda et al. [23,30], backpropagation neural networks outperform other methods in flood forecasting studies for their more efficiency and higher robustness. Berkhahn et al. [19] compared the training algorithms for hyperparameter tuning. The authors showed that resilient backpropagation is more efficient than both backpropagation and Levenberg–Marquardt for maximum flood inundation prediction. The process has two stages: the training stage gathers information from the flood event database, changing the weights between layers to minimize the error on the output layer; the recalling stage generates the forecast for the rest of the events in the database for testing the model.

Formula 1 and Formula 2 show the scheme of a resilient backpropagation. To calculate the update of the network weights w_{ij} from ith neuron to jth neuron, the gradient descent algorithm is applied. It distinguishes the update of weights upon the derivative of the loss function L of the model. The loss function L takes the mean square error (MSE). The iteration stops once the loss function reaches its minimum (chosen 10^{-6} in this case).

$$\Delta_{ij}(t) = \begin{cases} \eta^+ \cdot \Delta_{ij}(t-1), & \frac{\partial L}{\partial w_{ij}}(t) \cdot \frac{\partial L}{\partial w_{ij}}(t-1) > 0 \\ \eta^- \cdot \Delta_{ij}(t-1), & \frac{\partial L}{\partial w_{ij}}(t) \cdot \frac{\partial L}{\partial w_{ij}}(t-1) < 0 \\ \Delta_{ij}(t-1), & \text{else} \end{cases} \quad (1)$$

$$w_{ij}(t) = \begin{cases} w_{ij}(t-1) + \Delta_{ij}(t), & \frac{\partial L}{\partial w_{ij}}(t) < 0 \\ w_{ij}(t-1) - \Delta_{ij}(t), & \frac{\partial L}{\partial w_{ij}}(t) > 0 \\ 0, & \text{else} \end{cases} \quad (2)$$

The learning rate is to scale the speed in each weight updating iteration. The larger alternative learning rate η^+ is chosen when the error gradient in the same signal in neighboring iterations and lower alternative learning rate η^- when the loss function is close to zero, fulfilling $0 < \eta^- < 1 < \eta^+$. In our study, these were set constant and equal to $\eta^- = 0.5$, $\eta^+ = 1.2$. The deep learning toolbox of MATLAB version R2017a is used to form the forecasts.

2.3. Prediction of the First Interval of Flood Events

The ANN model is trained with the first 120 events in the synthetic flood event database (more details in Chapter 3.2). The time series of each event (starting from time 0) is extracted for training. The input inflow discharges are extracted from time 0 to X h (X takes the values from 3, 6, 9, 12). The respective output inundation maps at X h (X takes the values from 3, 6, 9, 12) are used as the output layer. The intervals of 3 h, 6 h, 9 h, 12 h of the flood events are used for training four networks with the same forecast lead times (see Figure 2). The ANN models only consider the input flow values from the initial time step (blue bars of the events in Figure 2), but not from the previous time steps. This was similar to the approach in the framework FloodEvac, which successfully produced forecasts based on the selection of pre-recorded flood maps [31].

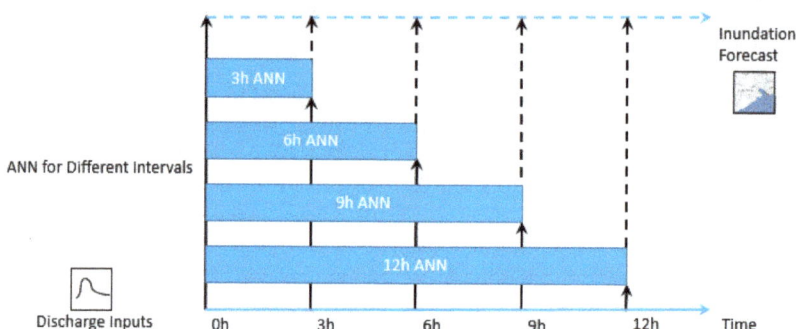

Figure 2. Training of artificial neural networks (ANN) forecast model. Four ANN models for 3 h, 6 h, 9 h, 12 h first interval predictions are set up in this work, trained with the discharges from each synthetic flood event. After this, the models are to predict the corresponding first intervals for other events.

After the training, the models are tested for the first interval forecast for the rest 60 events in the synthetic database.

2.4. Real-Time Forecasting for Sequential Multistep Forecast Intervals

In this work, the flood forecast starts when a certain discharge forecast threshold is achieved. If the start point occurs sometime later at time x, the prediction begin is also shifted to time x accordingly. If all the discharge inflows fall below the forecast threshold, the forecast is stopped. With this setup, the forecast can run in continuous mode.

The ANN receives the corresponding discharge inputs of an interval, just as in real-time forecasts. After the forecast is complete for a certain step, the discharge forecast is repeated one hour ahead. The forecasts are done with the same ANN model, now starting one hour later, taking the discharge inputs from the next time interval. This procedure is repeated many times to enable the continuous mode of flood forecasting. In this case study, the real-time forecast is performed with the ANN models trained to forecast at multiple steps of 1–5 h. The forecast from time 0 and the shift forward of the forecast intervals by one hour and two hours are shown in Figure 3.

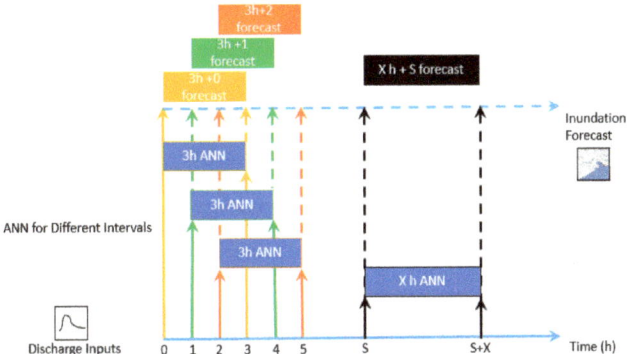

Figure 3. Shift of ANN forecast models for multistep forecast intervals. The yellow color shows the forecast of the first interval (forecast interval same as the training interval, i.e., at time 0). The green color shows applying the original 3 h forecast network for 1 h later forecast from 1–4 h. The orange color shows applying the original 3 h forecast network for 2 h later forecast from 2–5 h. The black box shows the general case of applying the original X h forecast network for S h later forecast, from S h to X h + S.

For an easier interpretation of the different forecast groups, we name each forecast as "X h + S" forecast. The "X h" indicates the forecast interval of X hours, and the "+ S" behind it shows the start time of the forecast.

2.5. Model Evaluation

The root-mean-square error (RMSE) is applied to access the ANN forecast performance in the study area. The forecasts of the ANN are compared against the inundation maps produced by the 2D dynamic model (see Section 3.2). Hence the 2D dynamic model results are assumed as the observed values in order to enable the evaluation of the ANN. Al, the events in the database have been processed by the FloodEvac tool [31] and validated [32]. As the ANN training is conducted within each grid, the RMSE is also evaluated for each grid.

$$\text{RMSE} = \sqrt{\frac{1}{n}\sum_{i=1}^{n}(T-S)^2}, \tag{3}$$

where

T is the predicted value, water depth from the ANN model in our case.
S is the observed value, water depth from the hydraulic model (HEC-RAS) in our case.
To assess the general conduct of the model over the training and validation dataset, the average RMSE is also calculated for the average accuracy among all the events in the testing dataset.

To quantify the forecast of inundation extent growth, the following indices are used to measure the correspondence between the ANN model and the hydraulic model, namely probability of detection (POD), false alarm ratio (FAR) and critical success index (CSI) [33].

$$POD = \frac{hits}{hits + misses}, \quad (4)$$

$$FAR = \frac{false\ alarms}{hits + false\ alarms}, \quad (5)$$

$$CSI = \frac{hits}{hits + misses + false\ alarms}, \quad (6)$$

A pixel with water depths under 10 cm is defined as a dry pixel, while over 10 cm as a wet pixel. Hits count the pixels that are both wet by the ANN forecast and the hydraulic simulation. Misses counting the pixels that are predicted dry by the ANN model but simulated wet by the hydraulic model. False alarms count the pixels predicted wet by ANN model but simulated dry by hydraulic model.

3. Study Area and Database

3.1. Study Area

The study area of Kulmbach lies by the River Main in Bavaria, Germany. The White Main divides the city to the north and south parts. Seven streams, specifically the Red Main, White Main, Dobrach, Schorgast, Mühlbach, Kohlenbach and Kinzelsbach, flow into this area. The city Kulmbach has a population of 25, 866 inhabitants in an area of 92.77 km². An extreme flood event hit the city on 28 May 2006. A flood mitigation plan was prepared by local stakeholders to mitigate future events. In the ANN model, the above seven streams are taken as the input boundary conditions. The goal of the ANN modeling is to replace the hydraulic processes within the marked study area to enable fast real-time forecasts (see Figure 4).

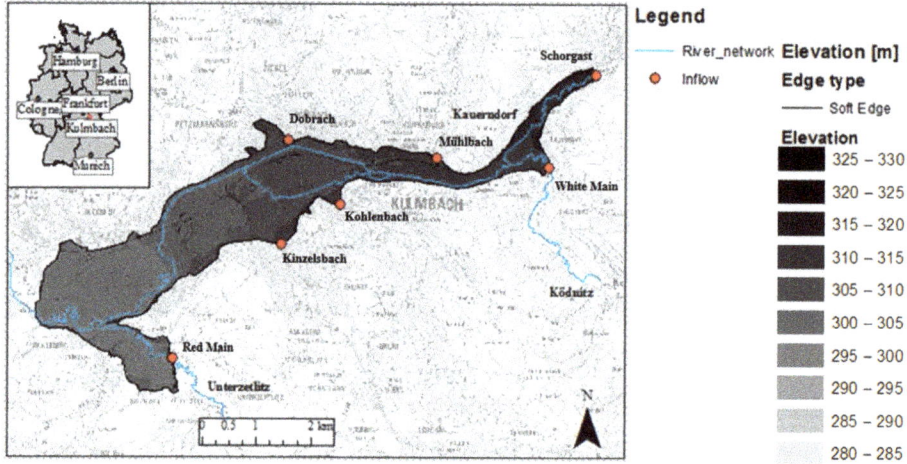

Figure 4. Map of the study area. It shows the location of Kulmbach in Germany. The blue curves represent the river network. The shaded region is the study area with its topography represented. On the marked boundary, the red points represent the seven inflows on the boundary (three rivers and four smaller streams).

3.2. HEC-RAS and Synthetic Event Database

The synthetic database is conducted by the 2D hydraulic model Hydrologic Engineering Center—river analysis system (HEC-RAS), Davis, CA, USA) for different precipitation durations, intensities and distributions [31]. Each event in the database contains a discharge hydrograph and an inundation map. The database contains 180 synthetic events in which discharge hydrographs are generated by the hydrologic model large area runoff simulation model (LARSIM) [34]. The events of the final database cover a wide range of different return periods, ranging from one year to 1.5 × 100 year return periods. The 2D hydrodynamic model HEC-RAS is used for producing the flood inundation maps. In the end, 180 hydrographs and their corresponding inundation maps form the synthetic event databases. The tool for automating these procedures is named the FloodEvac tool. The model is validated [32]. All the events are with a high temporal resolution of 15 min, and the inundation map is projected to a high spatial resolution (4 m by 4 m).

4. Results

The ANNs are trained for the first intervals (time 0). Thus, Section 4.1 focuses on assessing the results for the same time interval, which the ANNs are trained for. Section 4.2 focuses on the subsequent multistep forecast intervals. The aim is to verify if the hypothesis that the same ANNs can be used to forecast subsequent multistep intervals successfully even though they were trained for the first interval.

4.1. Assessment of the Prediction of Water Depths of the First Intervals (time 0) of Flood Events

4.1.1. Synthetic Flood Events

The ANN model is tested with the 60 synthetic flood events from the FloodEvac tool. The ANN model for the prediction of first intervals of flood events is set up for the duration of 3 h, 6 h, 9 h and 12 h, using the discharge within the same time as the model input. After this, the prediction of first intervals of flood events is evaluated by the RMSE with the testing dataset (event #121 to event #180). The averaged RMSE was calculated for different prediction times (3 h, 6 h, 9 h, 12 h) to quantify the prediction performance of each individual ANN. Table 1 shows the percentage of the accurate prediction ANNs, classified by RMSE of 0.2 m, 0.3 m and 0.4 m. In Table 1, if the error threshold is set to 0.3 m, the accuracy can be considered excellent with values above 80% for all the prediction durations.

Table 1. Number of wet grids and grid percentages of different large error thresholds for testing synthetic flood events (60 events, #121~#180).

Prediction Time (h)	Wet ANN Grid	ANN Grid with Average RMSE > 0.2 m	ANN Grid% with Average RMSE ≤ 0.2 m	ANN Grid with Average RMSE > 0.3 m	ANN Grid% with Average RMSE ≤ 0.3 m	ANN Grid with Average RMSE > 0.4 m	ANN Grid% with Average RMSE ≤ 0.4 m
3	300	47	84.33%	18	94.00%	10	96.67%
6	417	174	58.27%	78	81.29%	27	93.53%
9	474	106	77.64%	37	92.19%	15	96.84%
12	483	50	89.65%	12	97.52%	7	98.55%

Note: highlighted in gray are the percentages larger than 70%.

4.1.2. Historical Flood Events

After testing with the synthetic events, the ANN model performance is further examined with the historical flood events. Thus, the historical events, the same as the synthetic events, are simulated by the FloodEvac tool for their inundation maps. Afterward, the grid RMSE is calculated for the evaluation of prediction accuracy on the three historical flood events of their first intervals. A value of 10 m^3/s was selected as the forecast threshold to initiate the forecasts since this value is crossed before the beginning of the flooding in all three historical events. The forecast threshold is chosen slightly bigger than the average discharge of 9.2 m^3/s of White Main [35] to avoid the low discharges from triggering flood warnings.

Historical flood events 2006

Figure 5 shows the discharge inputs for the historical flood event in 2006. The first 3 h, 6 h, 9 h, 12 h discharge curves are given by the trained ANN as in Chapter 4.1. Figure 6 compares the prediction of the inundation map of the first intervals of 3 h, 6 h, 9 h and 12 h with the inundation map from the hydraulic model of the historical flood event 2006. Table 2 shows the performance of the prediction for historical event 2006, evaluated by average RMSE for each individual ANN. As the forecast interval increases from 3 h to 12 h, the prediction accuracy drops, evaluated by grid percentages of RMSE.

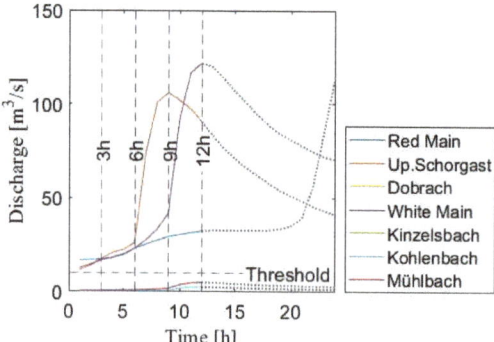

Figure 5. Hydrographs of the flood event in 2006. Seven discharge curves of three rivers and four streams are shown in different colors. Time 0 marks the start of the prediction. The dash lines upon the discharge curves mark the different discharge sections for prediction inputs.

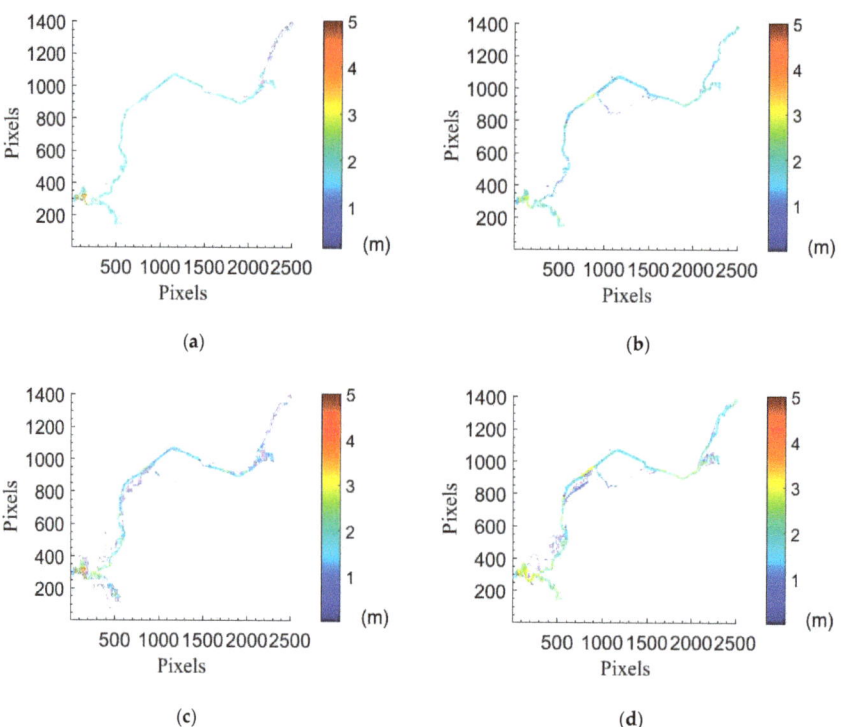

Figure 6. *Cont.*

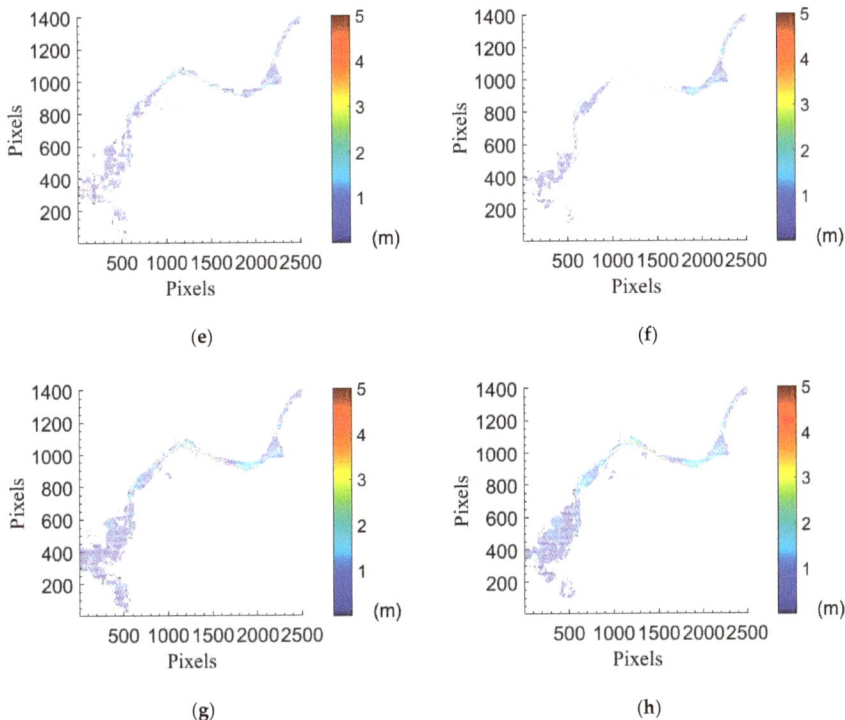

Figure 6. Inundation maps from the prediction of water depths of the first intervals in flood event 2006. (**a**) ANN inundation map 3 h; (**b**) hydrodynamic inundation map 3 h; (**c**) ANN inundation map 6 h; (**d**) hydrodynamic inundation map 6 h; (**e**) ANN inundation map 9 h; (**f**) hydrodynamic inundation map 9 h; (**g**) ANN inundation map 12 h; (**h**) hydrodynamic inundation map 12 h.

Table 2. Numbers of wet grids and accurate grid percentage for event 2006. A wet grid is with the water level over 0.1 m; any water depth below this cutoff value is eliminated. Table shows grid numbers with a larger root-mean-square error (RMSE) and their percentages to the total wet grids.

Prediction Time (h)	Wet ANN Grid	ANN Grid with RMSE > 0.2 m	ANN Grid% with RMSE ≤ 0.2 m	ANN Grid with RMSE > 0.3 m	ANN Grid% with RMSE ≤ 0.3 m	ANN Grid with RMSE > 0.4 m	ANN Grid% with RMSE ≤ 0.4 m
3	280	46	83.57%	20	92.86%	6	97.86%
6	405	84	79.26%	42	89.63%	25	93.83%
9	474	134	71.73%	64	86.50%	36	92.41%
12	483	157	67.49%	85	82.40%	47	90.27%

Note: highlighted in gray are the percentages larger than 70%.

Historical flood events 2013

Figure 7 shows the discharge inputs for the historical flood event in 2013. It shows that the initial discharge curves are below the forecast threshold 10 m^3/s; therefore, the start of the prediction at 9 h is marked with a red line when one discharge hits the forecast threshold. Figure 8 compares the prediction of the inundation map of the first intervals of 3 h, 6 h, 9 h and 12 h with the inundation map from the hydraulic model of the historical flood event 2013. Table 3 shows the performance of the prediction for historical event 2013, evaluated by average RMSE for each individual ANN. The forecast performance is slightly better than that of the event 2006.

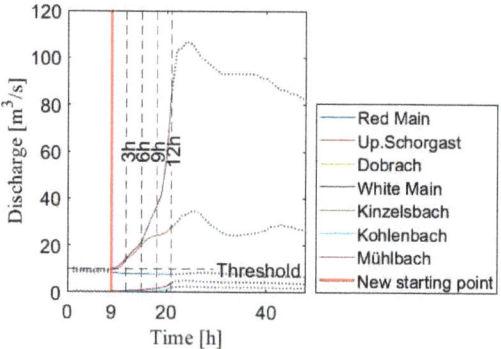

Figure 7. Hydrographs of the flood event in 2013. Seven discharge curves of three rivers and four streams are shown in different colors. The red line time marks the new start of the prediction at 9 h, where one discharge first exceeds the forecast threshold of 10 m^3/s. The dash lines upon the discharge curves mark the different discharge sections for prediction inputs.

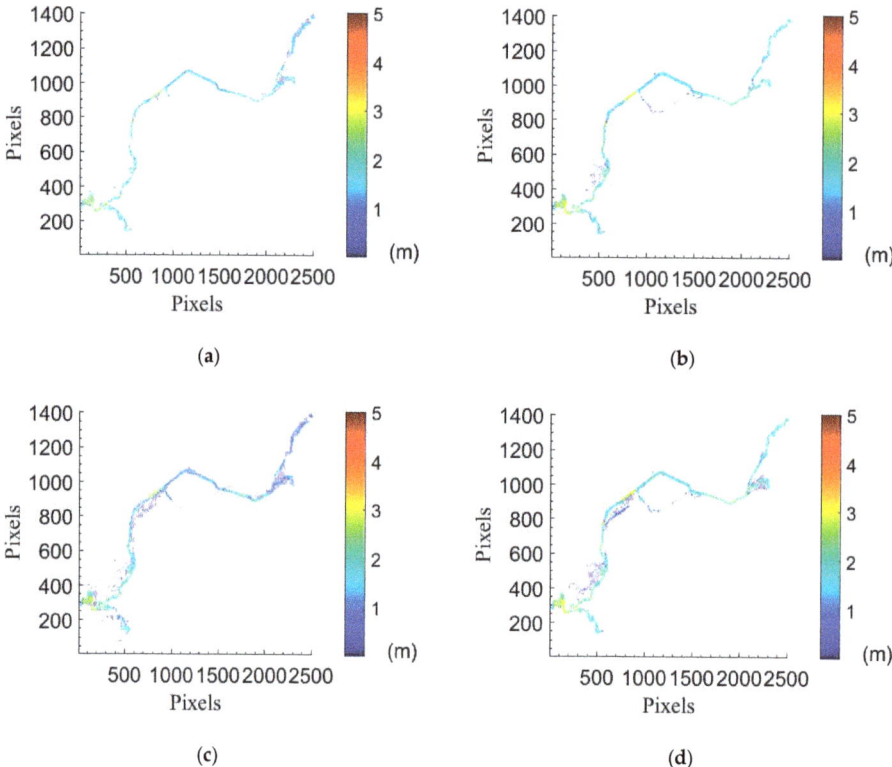

Figure 8. *Cont.*

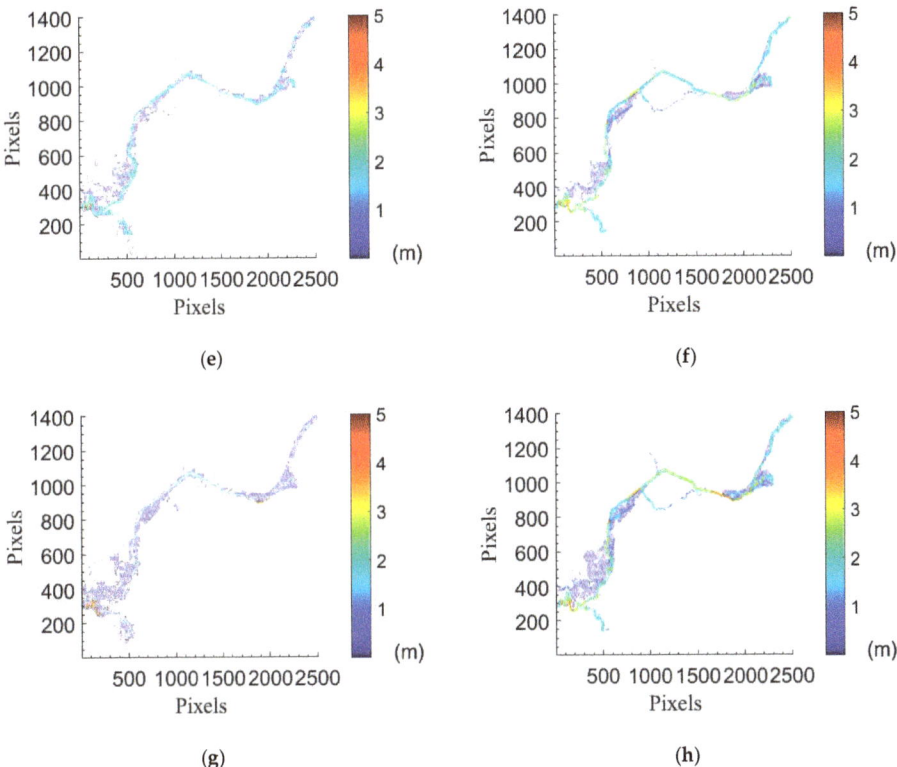

Figure 8. Inundation maps from the prediction of water depths of the first intervals in flood event 2013. (**a**) ANN inundation map 3 h; (**b**) hydrodynamic inundation map 3 h; (**c**) ANN inundation map 6 h; (**d**) hydrodynamic inundation map 6 h; (**e**) ANN inundation map 9 h; (**f**) hydrodynamic inundation map 9 h; (**g**) ANN inundation map 12 h; (**h**) hydrodynamic inundation map 12 h.

Table 3. Numbers of wet grids and accurate grid percentages for the flood event in 2013. A wet grid is with the water level over 0.1 m; any water depth below this cutoff value is eliminated. The table shows grid numbers with larger RMSE and their percentages to the total wet grids.

Prediction Time (h)	Wet ANN Grid	ANN Grid with RMSE > 0.2 m	ANN Grid% with RMSE ≤ 0.2 m	ANN Grid with RMSE > 0.3 m	ANN Grid% with RMSE ≤ 0.3 m	ANN Grid with RMSE > 0.4 m	ANN Grid% with RMSE ≤ 0.4 m
3	285	9	96.84%	2	99.30%	2	99.30%
6	405	72	82.22%	27	93.33%	8	98.02%
9	474	134	71.73%	65	86.29%	25	94.73%
12	483	175	63.77%	104	78.47%	56	88.41%

Note: highlighted in gray are the percentages larger than 70%.

Historical flood events 2005

Figure 9 shows the discharge inputs for the historical flood event in 2005. Figure 10 compares the prediction of the inundation map of the first intervals of 3 h, 6 h, 9 h and 12 h with the inundation map from the hydraulic model of the historical flood event 2005. Table 4 shows the prediction performance of historical event 2005, evaluated by average RMSE for each individual ANN. As the forecast interval increases from 3 h to 12 h, the prediction accuracy drops. This can be evaluated by the grid percentage of RMSE.

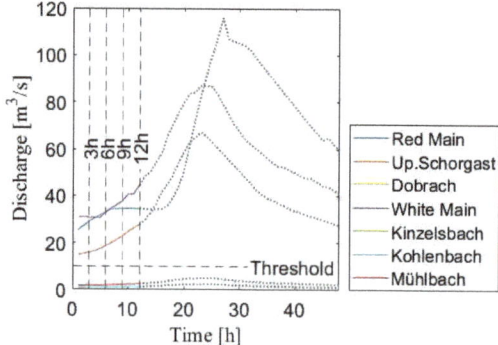

Figure 9. Hydrographs of the flood event in 2005. Seven discharge curves of three rivers and four streams are shown in different colors. Time 0 marks the start of the prediction. The dash lines upon the discharge curves mark the different discharge sections for prediction inputs.

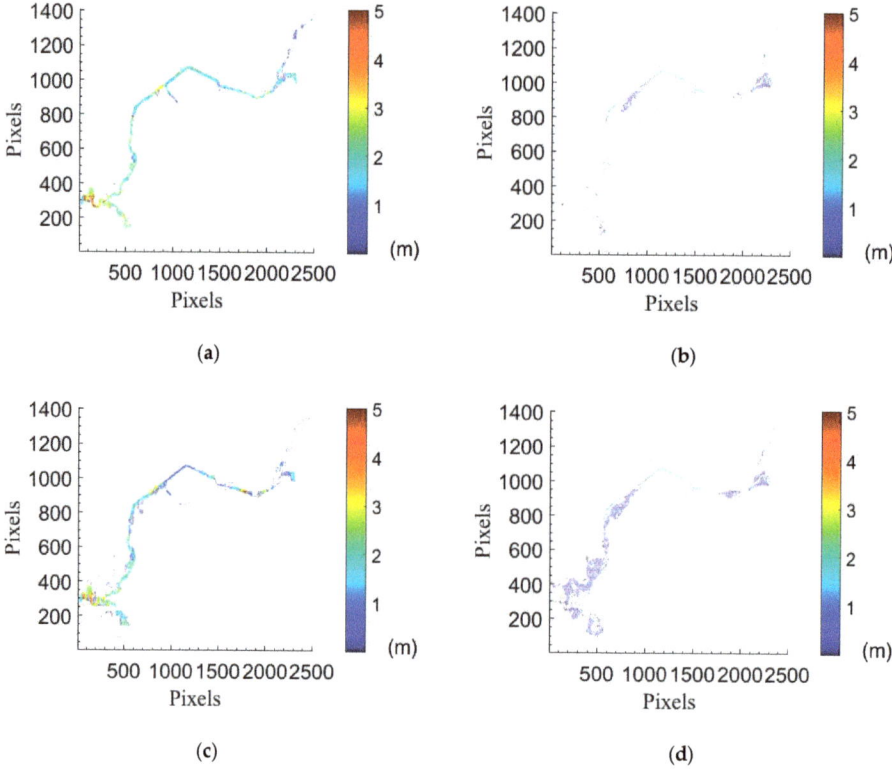

Figure 10. *Cont.*

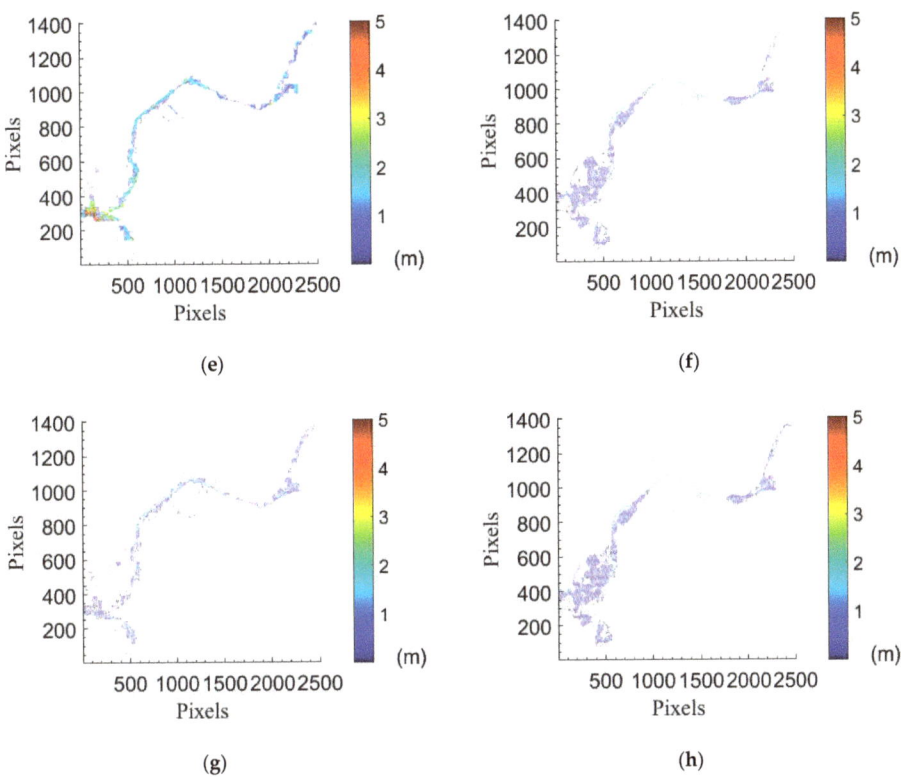

Figure 10. Inundation maps from the prediction of water depths of the first intervals in flood event 2005. (**a**) ANN inundation map 3 h; (**b**) hydrodynamic inundation map 3 h; (**c**) ANN inundation map 6 h; (**d**) hydrodynamic inundation map 6 h; (**e**) ANN inundation map 9 h; (**f**) hydrodynamic inundation map 9 h; (**g**) ANN inundation map 12 h; (**h**) hydrodynamic inundation map 12 h.

Table 4. Numbers of wet grids and accurate grid percentages for the flood event in 2005. A wet grid is with the water level over 0.1 m; any water depth below this cutoff value is eliminated. Table shows grid numbers with larger RMSE and their percentages to the total wet grids.

Prediction Time (h)	Wet ANN Grid	ANN Grid with RMSE > 0.2 m	ANN Grid% with RMSE ≤ 0.2 m	ANN Grid with RMSE > 0.3 m	ANN Grid% with RMSE ≤ 0.3 m	ANN Grid with RMSE > 0.4 m	ANN Grid% with RMSE ≤ 0.4 m
3	280	65	76.79%	36	87.14%	19	93.21%
6	405	165	59.26%	115	71.60%	74	81.73%
9	474	216	54.43%	148	68.78%	93	80.38%
12	483	244	49.48%	168	65.22%	107	77.85%

Note: highlighted in gray are the percentages larger than 70%.

4.2. Assessment of Real-time Forecasting of Water Depths for Multistep Flood Forecast Intervals, 1–5 h

Historical flood events 2006

Table 5 shows the forecast for multistep forecast intervals of the event in 2006. The forecast for the event in 2006 has good accuracy for all the intervals.

Table 5. Forecast accuracy percentages for the flood event in 2006. This table shows the grid percentage to the total wet grids with average RMSE within 0.3 m. The forecast begins by the starting point, several hours later than the event beginning for the real-time forecast.

Starting Point (h)	Prediction Interval (h)			
	3	6	9	12
+1	98.93%	92.10%	83.12%	79.09%
+2	98.94%	90.86%	77.85%	79.92%
+3	96.94%	89.38%	76.58%	78.26%
+4	95.11%	86.95%	70.89%	75.36%
+5	86.60%	69.29%	66.88%	68.12%

Note: highlighted in gray are the percentages larger than 70%.

Historical flood events 2013

Table 6 shows the forecast for multistep forecast intervals of the event in 2013. The forecast of the event in 2013 has good accuracy for all the intervals, with a similar performance as that of the event in 2006.

Table 6. Forecast accuracy percentages rate for the flood event in 2013. This table shows the grid percentage to the total wet grids with average RMSE within 0.3 m. The forecast begins by the starting point, several hours later than the event beginning for the real-time forecast.

Starting Point (h)	Prediction Interval (h)			
	3	6	9	12
+1	99.30%	92.59%	84.18%	72.67%
+2	98.95%	89.88%	78.90%	70.39%
+3	96.30%	89.17%	75.74%	67.91%
+4	94.17%	82.51%	68.78%	67.29%
+5	91.28%	77.34%	66.46%	66.67%

Note: highlighted in gray are the percentages larger than 70%.

Historical flood event 2005

Table 7 shows the forecast of multistep forecast intervals of the event 2005. The 3 h forecast of event 2005 still has a good accuracy of over 70%. For the 6 h, 9 h, 12 h forecasts, the ANN model produces less accurate results.

Table 7. Forecast accuracy percentages for the flood event in 2005. This table shows the grid percentages to the total wet grids with average RMSE within 0.3 m. The forecast begins by the starting point, several hours later than the event beginning for the real-time forecast.

Starting Point (h)	Prediction Interval (h)			
	3	6	9	12
+1	83.74%	69.95%	66.89%	63.15%
+2	81.67%	68.23%	64.77%	61.70%
+3	76.31%	67.00%	63.08%	60.25%
+4	74.85%	66.50%	60.97%	60.25%
+5	70.97%	61.08%	58.65%	60.25%

Note: highlighted in gray are the percentages larger than 70%.

4.3. Forecast of the Inundation Extent

Figure 11 evaluates the ANN performance of the forecast of inundation extent (water depths over 0.1 m) growth with three indices. The status of wet/dry is used for calculating the following three indices for the likelihood between ANN and the hydraulic model. The probability of detection (POD) represents how well the ANN forecasted the same inundation extent as the hydraulic model. The false alarm ratio (FAR) measures the discrepancy of the ANN forecast to the hydraulic model. The critical success index (CSI) is the ratio between the correct forecasted inundation and the join of both the inundations (hits + misses + false alarms), showing the general correctness of the flood extent forecast of the ANN model. According to the verification criteria in another study [36], CSI over 0.7 (see Figure 11) is considered a good fit for the benchmark and over 0.5 (see Figure 11) is a sufficient fit. The lines in the figures show how these three indices change when a specific ANN is performed as the multistep forecasts advance in time.

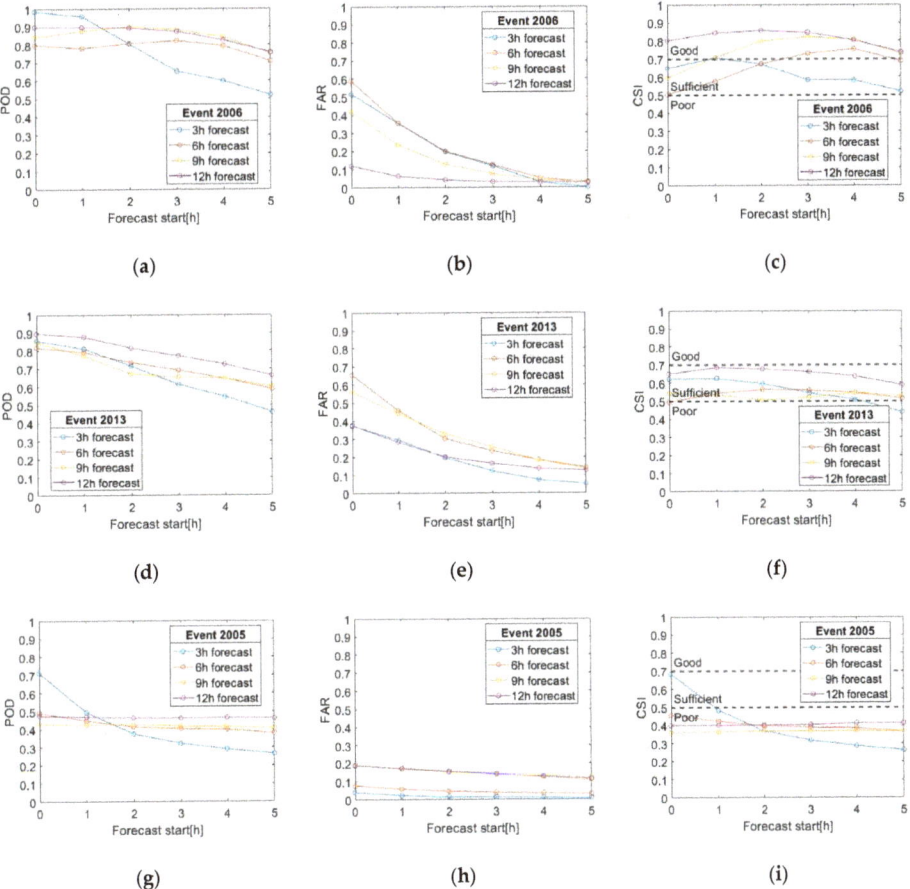

Figure 11. Performance of the forecast of inundation extent growths by three indices. (**a**) probability of detection (POD) in flood event 2006; (**b**) false alarm ratio (FAR) in flood event 2006; (**c**) critical success index (CSI) in the flood event 2006; (**d**) POD in flood event 2013; (**e**) FAR in the flood event 2013; (**f**) CSI in flood event 2013; (**g**) POD in flood event 2005; (**h**) FAR in flood event 2005; (**i**) CSI in flood event 2005.

5. Discussion

5.1. Assessment of the Prediction of Water Depths of the First Intervals of Flood Events

5.1.1. Synthetic Flood Events

Overall, the performance of the prediction of the first intervals of flood events shows that more than 80% of the grids have errors smaller than 0.3 m. The accuracy table (see Table 1) shows that the ANN has good accuracy in the water depth prediction of first intervals for all the synthetic events in the testing dataset. This test validated the network structure as well as the resilient backpropagation for solving the ANN. Since the network is initially trained with the 120 events from the synthetic event database and validated with the rest 60 events. The training events and the validation events bear more similarities with each other, which could explain the good performance of the prediction of the first interval of the ANN.

5.1.2. Historical Flood Events

Historical flood event 2006

From Table 2, 83% of grids have RMSE smaller than 0.2 m, and the rest of the grids have RMSE around zero in 3 h prediction. For the 6 h prediction, 79% of grids have RMSE less than 0.2 m. In the 9 h and 12 h prediction, the area with large errors grows slightly (see Table 2). From Figure 6, the inundation maps from 3 h and 6 h predictions match well with the hydraulic inundation maps. The 9 h and 12 h are less accurate, especially in the southwest of the study area, which is further away from the location of the discharge inflows and is, thus, likely less sensitive to the changes in the discharge inputs. In brief, all the prediction for flood event 2006 is precise with more than 82% grids having RMSE less than 0.3 m.

Historical flood event 2013

For the historical flood event 2013, the discharge forecast threshold for the forecast start was reached later, signaling that the discharge forecast threshold is indeed effective in starting and stopping the forecast. Therefore, the start of the forecast is picked up at a later moment in time once one discharge crosses the forecast threshold of 10 m^3/s for a second time. The red line in Figure 7 marks the new start of the forecast (red line). For this event, it is nine hours later after the first forecast signaled by the ANN. Table 3 shows that the ANN model achieved high accuracy for the flood event in 2013. For the 3 h prediction, 96% of the grids have RMSE less than 0.2 m. 6 h prediction has 82% grids with RMSE less than 0.2 m. From 3 h and 6 h prediction of event 2013, the ANN performs better than the event 2006. Overall, the event of 2013 is also well predicted, with over 78% of grids having RMSE less than 0.3 m. Similar to event 2006, the predicted flood inundation maps of 3 h and 6 h intervals are similar to the hydraulic inundation simulations (see Figure 10).

Historical flood event 2005

The general model performance for this flood event is less good. Figure 10 shows the comparison between the predicted inundation maps of ANN, where the inundated area is underestimated, compared to the hydraulic model (see the dark blue area in Figure 10b,d,f,h. However, Table 4 shows the prediction accuracy is still sufficient (above 65%) considering as acceptable an error within 0.3 m for the prediction of the first intervals for flood event 2005. The comparison can be observed in the water depth maps in Figures 6, 8 and 10, particularly when comparing the subplots (e) with (f) and (g) with (h). The ANN results differ more from the hydraulic results at points located closer to the southwest end of the study area. Reason being that those are the points that are further away from the inflow points (Figure 4).

5.2. Assessment of Real-time Forecasting of Water Depths for Multistep Flood Forecast Intervals, 1–5 h

In this section, we investigate if we can train different ANNs for the first interval of the forecast (from time 0) to predict the multistep forecast, the hypothesis of this study. Herein we test the ANN for four forecast intervals, namely for the next 3, 6, 9, 12 hours (X), and for five forecast starts 1–5 hours (S). This is represented by the format "X h + S".

Historical flood event 2006

Table 5 shows the forecast accuracy of the historical flood event in 2006. The forecast of the 2006 flood event shows a good accuracy in 3 h, 6 h and most of 9 h (over 70% grid with RMSE < 0.3 m). It is visible that as the multistep forecast shifts further away from the original start used for the model training (X h + 5 h forecasts in Table 5), the ANN model performance decreases.

Historical flood event 2013

The discharge forecast threshold of the flood event from 2013 exceeds shortly at the beginning. Hence, the forecast is deactivated and reactivated again when the discharge exceeds the forecast threshold of 10 m^3/s for the second time, namely 9 hours after time 0 (the first time the forecast window was activated). From the second starting point, all other forecasts are done for every forecast for X h + 1 h to X h + 5 h. Table 6 shows the forecast accuracy of the historical flood event of 2013. From this table, the ANN model performs as similar to the flood event in 2006. The forecast of the 2013 flood event has good results in 3 h, 6 h and most of 9 h (over 70% of the grid with RMSE < 0.3 m).

Historical flood event 2005

Table 7 shows the accuracy percentage of the grids evaluated by the RMSE less than 0.3 m. With the changing of the forecast starting point, all the forecasts of different intervals have similar RMSE as the forecast done during the first intervals. It is noticeable that the model provides a good forecast (over 70% grids with RMSE < 0.3 m) for 3 h intervals for all starting points. This shows that 3 h ANN trained for the first interval could be used to forecast subsequent intervals with a slight drop in the overall accuracy. However, the forecasts of 6 h, 9 h and 12 h show a poor performance (Tables 5–7). Similar to the other events, most of the errors occur in the southwest of the study area (Figure 10). However, in this particular event, the errors are substantially larger at the southwest than at the city center, hence the overall poor performance of the ANNs.

In all the three historical events, the forecast accuracy decreases as the forecast interval increases from 3 h to 12 h (see Tables 5–7). One exception occurs in the event 2006 between 9 h and 12 h, where the 12 h forecast has higher accuracy than the 9 h forecast. From the discharge curve (see Figure 5), unlike in other events, the two major discharges are falling after the peak value, which could be the reason for the higher accuracy at 12 h in this case.

5.3. Forecast of the Inundation Extent

The forecasts of flood inundation extent growths are examined through the statistical analysis proposed by Li et al. [33]. Figure 11 shows three indices, POD, FAR and CSI, for measuring the forecast performance of the flood inundation extent. Analyzing the POD index (see Figure 11a,d,g), it is clear that for the 3 h ANN forecast, the accuracy decreases slightly as forecasts proceeds from 3 h + 0 h to 3 h + 5 h. In other words, the accuracy of the 3 h ANN network is more sensitive to the shift of the forecast intervals than the 6 h, 9 h and 12 h ANN networks. The 3 h network achieves the best forecast performance for the first interval (training interval same as the forecast intervals). When moving forward for multistep forecast, shifting each hour decreases the POD by a value that varies between 0.08 to 0.1. This means that an added 8% to 10% of the inundation extent displayed by the hydraulic model is missing in the ANN forecast. In any case, and except for the event of 2005, the POD exhibits values above 70% for the first 2 hours of the forecast.

The FAR index (see Figure 11b,e,h) indicates the false-alarm percentage of the ANN forecasted flood inundation extents. In all three events, it is noticeable that the area percentage with false-alarms decreases over all the forecast networks when the forecast interval moves forward. It shows the ANN forecast produces more percentage of false alarms at the early stage in a flood event. This is because the flood inundation is relatively small at the beginning and the number of outer pixels larger than that of inner pixels, causing a higher number of false alarms. Moreover, the decreasing trend of POD and FAR indicate that the ANN model tends to change from overestimation to underestimation when the forecast starts to shift from 0–5 h.

The CSI index (see Figure 11c,f,i) shows the percentage of agreement of the ANN forecasts of the flood inundation extent to the hydraulic model. The ANN predicts better the flood inundation extents for two events of 2006 and 2013 than for the event of 2005, with the CSIs from 2006 and 2013 close to 0.6. For the event of 2005, the CSI is around 0.4 showing a poor accuracy forecast in the flood inundation extent forecast.

It is noteworthy to mention that the ANN shows an expectable performance of water depth prediction decreasing with lead time (see Tables 5–7). However, if we focus on the inundation extent, it seems contradictory, as the 12 h prediction shows better performance (see CSI and POD in Figure 11). The latter apparent contraction can be explained by the flood inundation extent being limited by the topography. The topography limits the size of the inundation, making it easier for the ANN to predict it better.

6. Conclusions

The aim of this study is to perform multiple subsequent forecasts for 1–5 h after the flooding event has started. It was shown that it is possible to use different ANNs for the first interval of the forecast (time 0) to issue the multistep forecasts. However, there should be made a distinction between the quality of the forecast regarding the water-depths or the flood inundation extent. The overall forecast performance of the water-depths was found slightly better than the flood inundation extents. The performance was mostly adversity affected by the flood event from 2005, in particular, close to the southwest end, far away from the location where the input inflows are.

The ANN model was first applied to the forecast of the first intervals of 3 h, 6 h, 9 h and 12 h. For the 60 synthetic flood events in the testing dataset, the model produced good results, as over 81% grids with RMSE less than 0.3 m. For the historical event 2006 and the historical event 2013, the model performed good water depths with the accuracy of over 82% and 78%, evaluated by RMSE smaller than the error threshold of 0.3 m. The flood event 2005 has a sufficient performance with an accuracy of over 65%, evaluated by RMSE smaller than 0.3 m. The forecasted inundation maps by ANN of all the three historical events have a similar shape to the inundation maps from the hydrodynamic model (HEC-RAS). For the far end area away from the inflow inputs, the long-distance may be responsible for a decrease in the forecast performance; therefore, it is likely that the model requires other information than those of discharge to enhance the forecast accuracy for those areas.

The ANN model was applied for the real-time forecast of the historical events in 2006, 2013 and 2005. For this purpose, the same ANN model was used for the forecast. The input discharge inputs were replaced by the shifted intervals for 1–5 h after the event's beginnings. The forecast shows good results in the flood events 2006 and 2013 for the real-time forecasts, with over 70% grids with RMSE less than 0.3 m. The forecast shows worse results in flood event 2005, with only over 58% grids with RMSE less than 0.3 m. Overall, the forecast accuracy drops as the forecast interval increased from 3 h to 12 h. The forecast accuracy also decreases as the forecast progresses forward from X h + 1 h to X h + 5 h. For all the three historical flood events, the 3 h forecast is classified as good, with more than 70% grids accurately forecasted. However, the quality of 6 h or longer intervals was more event dependent.

Based on the analysis of indices of POD, FAR and CSI, the multistep ANN flood forecast provides good results at the beginning and decreases as the forecast progresses. The forecasts of the ANN model switches from an overestimation to an underestimation when the forecast proceeds from 0 h to 5 h. In our case, except for the event 2005, the 3 h ANN trained by the first interval improved the performances slightly with the multistep forecast; the 6 h, 9 h and 12 h ANN trained by the first interval for multistep interval forecasts would have accuracies depending on the exact flood events.

Future research could include recurrent neural networks with long short-term memory to involve the water depth information acquired from previous forecasted steps for a multistep forecast. To reduce the forecasted time interval for finer temporal multisteps could also be another possibility to enhance the accuracy of the forecast.

Author Contributions: Conceptualization, J.L.; methodology, Q.L.; data curation, S.G.; software, S.G.; investigation, Q.L.; visualization, Q.L.; writing—original draft preparation, Q.L.; writing—review and editing, Q.L., J.L. and M.D.; formal analysis, Q.L.; validation, Q.L. and S.G.; supervision, M.D. All authors have read and agreed to the published version of the manuscript.

Funding: This research was funded by Bavarian State Ministry of the Environment and Consumer Protection (StMUV) with the grant number 69-0270-50433/2017. The APC was funded by Technical University of Munich.

Acknowledgments: The research presented in this paper has been carried out as part of the HiOS project (Hinweiskarte Oberflächenabfluss und Sturzflut) funded by the Bavarian State Ministry of the Environment and Consumer Protection (StMUV) and supervised by the Bavarian Environment Agency (LfU).

Conflicts of Interest: The authors declare that they have no known competing financial interests or personal relationships that could have appeared to influence the work reported in this paper.

References

1. Berz, G. Flood disasters: Lessons from the past-worries for the future. *Water Manag.* **2001**, *148*, 57–58. [CrossRef]
2. Chang, L.C.; Amin, M.Z.M.; Yang, S.N.; Chang, F.J. Building ANN-based regional multi-step-ahead flood inundation forecast models. *Water* **2018**, *10*, 1283. [CrossRef]
3. Henonin, J.; Russo, B.; Mark, O.; Gourbesville, P. Real-time urban flood forecasting and modelling—A state of the art. *J. Hydroinform.* **2013**, *15*, 717–736. [CrossRef]
4. Hankin, B.; Waller, S.; Astle, G.; Kellagher, R. Mapping space for water: Screening for urban flash flooding. *J. Flood Risk Manag.* **2008**, *1*, 13–22. [CrossRef]
5. Kalyanapu, A.J.; Shankar, S.; Pardyjak, E.R.; Judi, D.R.; Burian, S.J. Assessment of GPU computational enhancement to a 2D flood model. *Environ. Model. Softw.* **2011**, *26*, 1009–1016. [CrossRef]
6. Mosavi, A.; Ozturk, P.; Chau, K.W. Flood prediction using machine learning models: Literature review. *Water* **2018**, *10*, 1–40.
7. Dineva, A.; Várkonyi-Kóczy, A.R.; Tar, J.K. Fuzzy expert system for automatic wavelet shrinkage procedure selection for noise suppression. In Proceedings of the INES 2014 IEEE 18th Internationl Conference Intelligent Engineering Systems, Tihany, Hungary, 3–5 July 2014; pp. 163–168.
8. Bermúdez, M.; Cea, L.; Puertas, J. A rapid flood inundation model for hazard mapping based on least squares support vector machine regression. *J. Flood Risk Manag.* **2019**, *12*, 1–14. [CrossRef]
9. Gizaw, M.S.; Gan, T.Y. Regional Flood Frequency Analysis using Support Vector Regression under historical and future climate. *J. Hydrol.* **2016**, *538*, 387–398.
10. Taherei Ghazvinei, P.; Darvishi, H.H.; Mosavi, A.; Bin Wan Yusof, K.; Alizamir, M.; Shamshirband, S.; Chau, K.W. Sugarcane growth prediction based on meteorological parameters using extreme learning machine and artificial neural network. *Eng. Appl. Comput. Fluid Mech.* **2018**, *12*, 738–749. [CrossRef]
11. Noymanee, J.; Nikitin, N.O.; Kalyuzhnaya, A.V. Urban Pluvial Flood Forecasting using Open Data with Machine Learning Techniques in Pattani Basin. *Procedia Comput. Sci.* **2017**, *119*, 288–297. [CrossRef]
12. Kasiviswanathan, K.S.; He, J.; Sudheer, K.P.; Tay, J.H. Potential application of wavelet neural network ensemble to forecast streamflow for flood management. *J. Hydrol.* **2016**, *536*, 161–173. [CrossRef]
13. Zhang, G.P. Avoiding pitfalls in neural network research. *IEEE Trans. Syst. Man Cybern. Part C Appl. Rev.* **2007**, *37*, 3–16. [CrossRef]
14. Dawson, C.W.; Wilby, R.L. Hydrological modelling using artificial neural networks. *Prog. Phys. Geogr.* **2001**, *25*, 80–108. [CrossRef]
15. Yu, P.S.; Chen, S.T.; Chang, I.F. Support vector regression for real-time flood stage forecasting. *J. Hydrol.* **2006**, *328*, 704–716.

16. Sit, M.; Demir, I. Decentralized flood forecasting using deep neural networks. *arXiv* **2019**, arXiv:1902.02308.
17. Bustami, R.; Bessaih, N.; Bong, C.; Suhaili, S. Artificial neural network for precipitation and water level predictions of bedup River. *IAENG Int. J. Comput. Sci.* **2007**, *34*, 228–233.
18. French, J.; Mawdsley, R.; Fujiyama, T.; Achuthan, K. Combining machine learning with computational hydrodynamics for prediction of tidal surge inundation at estuarine ports. *Procedia IUTAM* **2017**, *25*, 28–35. [CrossRef]
19. Berkhahn, S.; Fuchs, L.; Neuweiler, I. An ensemble neural network model for real-time prediction of urban floods. *J. Hydrol.* **2019**, *575*, 743–754. [CrossRef]
20. Lin, Q.; Leandro, J.; Wu, W.; Bhola, P.; Disse, M. Prediction of Maximum Flood Inundation Extents with Resilient Backpropagation Neural Network: Case Study of Kulmbach. *Front. Earth Sci.* **2020**, *8*, 1–8.
21. Shen, H.-Y.; Chang, L.-C. On-line multistep-ahead inundation depth forecasts by recurrent NARX networks. *Hydrol. Earth Syst. Sci. Discuss.* **2012**, *9*, 11999–12028. [CrossRef]
22. Sit, M.; Demiray, B.Z.; Xiang, Z.; Ewing, G.J.; Sermet, Y.; Demir, I. A comprehensive review of deep learning applications in hydrology and water resources. *Water Sci. Technol.* **2020**. [CrossRef]
23. Nawi, N.; Ransing, R.; Ransing, M. An improved Conjugate Gradient based learning algorithm for back propagation neural networks. *Int. J. Comput. Intell.* **2007**, *4*, 46–55.
24. Sankaranarayanan, S.; Prabhakar, M.; Satish, S.; Jain, P.; Ramprasad, A.; Krishnan, A. Flood prediction based on weather parameters using deep learning. *J. Water Clim. Chang.* **2019**, 1–18. [CrossRef]
25. Kohavi, R. A study of cross-validation and bootstrap for accuracy estimation and model selection. In Proceedings of the Appears in the International Joint Conference on Articial Intelligence (IJCAI), Stanford, CA, USA, 19 August 1995; Volume 8.
26. Jhong, Y.D.; Chen, C.S.; Lin, H.P.; Chen, S.T. Physical hybrid neural network model to forecast typhoon floods. *Water* **2018**, *10*, 632. [CrossRef]
27. Wang, H.; Rahnamayan, S.; Wu, Z. Parallel differential evolution with self-adapting control parameters and generalized opposition-based learning for solving high-dimensional optimization problems. *J. Parallel Distrib. Comput.* **2013**, *73*, 62–73. [CrossRef]
28. Kabir, S.; Patidar, S.; Xia, X.; Liang, Q.; Neal, J.; Pender, G. A deep convolutional neural network model for rapid prediction of fluvial flood inundation. *J. Hydrol.* **2020**, *590*, 125481. [CrossRef]
29. Saini, L.M. Peak load forecasting using Bayesian regularization, Resilient and adaptive backpropagation learning based artificial neural networks. *Electr. Power Syst. Res.* **2008**, *78*, 1302–1310. [CrossRef]
30. Panda, R.K.; Pramanik, N.; Bala, B. Simulation of river stage using artificial neural network and MIKE 11 hydrodynamic model. *Comput. Geosci.* **2010**, *36*, 735–745. [CrossRef]
31. Bhola, P.K.; Leandro, J.; Disse, M. Framework for offline flood inundation forecasts for two-dimensional hydrodynamic models. *Geosciences* **2018**, *8*, 346. [CrossRef]
32. Bhola, P.K.; Nair, B.B.; Leandro, J.; Rao, S.N.; Disse, M. Flood inundation forecasts using validation data generated with the assistance of computer vision. *J. Hydroinform.* **2019**, *21*, 240–256. [CrossRef]
33. Li, L.; Hong, Y.; Wang, J.; Adler, R.F.; Policelli, F.S.; Habib, S.; Irwn, D.; Korme, T.; Okello, L. Evaluation of the real-time TRMM-based multi-satellite precipitation analysis for an operational flood prediction system in Nzoia Basin, Lake Victoria, Africa. *Nat. Hazards* **2009**, *50*, 109–123. [CrossRef]
34. Ludwig, K. The Water Balance Model LARSIM: Design, Content and Applications, Freiburger Schriften zur Hydrologie. Master's Thesis, Institut für Hydrologie der University, Hannover, Germany, 2006.
35. Hochwassernachrichtendienst Bayern, added level data (MQ) from water level Ködnitz/White Main. Available online: www.hnd.bayern.de (accessed on 12 December 2019).
36. Bernhofen, M.V.; Whyman, C.; Trigg, M.A.; Sleigh, P.A.; Smith, A.M.; Sampson, C.C.; Yamazaki, D.; Ward, P.J.; Rudari, R.; Pappenberger, F.; et al. A first collective validation of global fluvial flood models for major floods in Nigeria and Mozambique. *Environ. Res. Lett.* **2018**, *13*. [CrossRef]

Publisher's Note: MDPI stays neutral with regard to jurisdictional claims in published maps and institutional affiliations.

© 2020 by the authors. Licensee MDPI, Basel, Switzerland. This article is an open access article distributed under the terms and conditions of the Creative Commons Attribution (CC BY) license (http://creativecommons.org/licenses/by/4.0/).

Article

Urbanization and Floods in Sub-Saharan Africa: Spatiotemporal Study and Analysis of Vulnerability Factors—Case of Antananarivo Agglomeration (Madagascar)

Fenosoa Nantenaina Ramiaramanana *[ ] and Jacques Teller [ ]

Local Environment Management and Analysis (LEMA), Urban & Environmental Engineering Department, University of Liège, 4000 Liège, Belgium; Jacques.Teller@uliege.be
* Correspondence: Fenosoa.Ramiaramanana@student.uliege.be

Citation: Ramiaramanana, F.N.; Teller, J. Urbanization and Floods in Sub-Saharan Africa: Spatiotemporal Study and Analysis of Vulnerability Factors—Case of Antananarivo Agglomeration (Madagascar). *Water* **2021**, *13*, 149. https://doi.org/10.3390/w13020149

Received: 29 October 2020
Accepted: 7 January 2021
Published: 10 January 2021

Publisher's Note: MDPI stays neutral with regard to jurisdictional claims in published maps and institutional affiliations.

Copyright: © 2021 by the authors. Licensee MDPI, Basel, Switzerland. This article is an open access article distributed under the terms and conditions of the Creative Commons Attribution (CC BY) license (https://creativecommons.org/licenses/by/4.0/).

Abstract: Flooding is currently one of the major threats to cities in Sub-Saharan Africa (SSA). The demographic change caused by the high rate of natural increase, combined with the migration toward cities, leads to a strong demand for housing and promotes urbanization. Given the insufficiency or absence of adequate planning, many constructions are installed in flood-prone zones, often without adequate infrastructure, especially drainage systems. This makes them very vulnerable. Our research consists of carrying out a spatiotemporal analysis of the agglomeration of Antananarivo (Madagascar). It shows that urbanization leads to increased exposure of populations and constructions to floods. There is a pressure on land in flood-prone zones due to the exponential growth of the population at the agglomeration level. Some 32% of the population of the Antananarivo agglomeration lived in flood-prone zones in 2018. An analysis of the evolution of built spaces from 1953 to 2017 highlights that urban expansion was intense over those years (6.1% yearly increase of built areas). This expansion triggered the construction of built areas in flood-prone zones, which evolved from 399 ha in 1953 to 3675 ha in 2017. In 2017, 23% of the buildings in the agglomeration, i.e., almost one out of every four buildings, were in flood-prone zones. A share of the urban expansion in flood-prone zones is related to informal developments that gather highly vulnerable groups with very little in terms of economic resources. Better integration of flood risk management in spatial planning policies thus appears to be an essential step to guide decisions so as to coordinate the development of urban areas and drainage networks in a sustainable way, considering the vulnerability of the population living in the most exposed areas.

Keywords: demographic change; urbanization; flooding; drainage system; vulnerability; Sub-Saharan Africa; Antananarivo

1. Introduction

Floods have become more recurrent and usual events in several countries [1]. Compared to the figures of the 1990s, the number of floods has almost doubled in the world since the 2000s [2]. They represent a threat with a major impact in terms of victims. In 2018, floods accounted for 50% of people affected by natural hazards [3]. Floods also have severe consequences in terms of economic loss and material damage [4].

The upsurge in floods can be explained by various factors, including climate change [5], which generates changes in precipitation regimes and intensity [6], and often manifests in torrential rains. Intense precipitation can cause flooding in small river basins and in rivers [4]. Extreme events in Africa [7], Europe [8], and Asia [9] are examples of the significance of climate change in increasing flooding. However, floods are not exclusively linked to climate change, but also to urbanization dynamics [10]. Jha, Bloch, and Lamond argue that regardless of climate change, urbanization can increase the risk of flooding [11]. With an emphasis on exposure and vulnerability, we would like to highlight this aspect in this paper.

Urbanization generally leads to an increase in impervious surfaces, which limits the possibility of water infiltration in the soil and increases the volume of water runoff on the surface [12]. Additionally, urbanization is often accompanied by an artificialization of urban rivers, which further increases the risk of water overflows [13,14]. This modifies existing land use not only inside cities but also in the outskirts [15].

In 1900, 15% of the world's population lived in urban areas [16]. Currently the proportion is more than 50% [17]. The numbers are increasing by 200,000 people a day, or 70 million people a year, and the proportion is estimated to reach 70% in 2050 [17]. This urban growth increases the demand for housing and land to build [16]. Given the competition for urban land, some people are tempted to build on areas exposed to risks [18,19].

Controlling exposure to floods implies a combination of urban planning and management of drainage systems. It requires follow-up of spatial planning policies [11], because risks are partly related to governance [1]. A lack of planning or poor planning can lead to an increase of informal installations and constructions, often exposing vulnerable residents to risks [10,20]. Lower-income residents usually do not have access to services and infrastructures that could mitigate the problems [21]. On the other hand, extending the drainage system should go hand-in-hand with any increase in built spaces, and it should be resilient [22] by having the capacity to evacuate water in the face of flooding. Without an adequate, sufficient, and well-maintained drainage system [23], urbanization cannot be sustainable.

Africa is one of the two continents in the world most affected by floods [24]. Floods are the most frequent disaster and remain a threat, especially in Sub-Saharan Africa (SSA)'s cities [25,26]. At the same time, the continent contains a population that is growing twice as fast other regions in the world [27]. Beyond this high growth, management and planning remains a problem throughout the continent and particularly in the region south of the Sahara [23,28]. The absence or ineffectiveness of disaster management plans and the inadequacy of basic systems, infrastructures, and services contribute to increasing vulnerability of urban areas [11]. The inability to accommodate a fast-growing population in decent conditions explains why constructions are located on unsuitable and dangerous sites, exposing cities to natural disasters [18], including floods. The deficiency of the drainage systems means a part of the population is affected by floods [14].

In this study, we show that urbanization leads to increased exposure of populations and constructions to floods and tends to add to their vulnerability. In order to reduce exposure and vulnerability to floods, it is important to recognize all aspects related to flooding, including socioeconomic factors that explain why flood-prone zones keep attracting a part of the population. We thus adopt a co-evolutionary perspective in order to better understand the long-term bi-directional relations between flood exposure, urban expansion, and vulnerability [29,30].

This paper is centered on the agglomeration of Antananarivo, the capital of Madagascar. Apart from the extreme climatic hazards the country is exposed to every year [31], it has most of the characteristics of SSA's cities mentioned above, in particular growing urbanization. According to the Institut National de la Statistique de Madagascar (INSTAT), nearly 5 million Malagasy people lived in urban areas in 2018 [32]. Rapid urban expansion is a problem due to the lack of planning [33]. It is associated with drainage problems plaguing the country. Insufficient capacity and poor functioning of the drainage network due to clogging with solid waste and deterioration are among the causes of floods [34]. The growing urban population is settling more and more in flood-prone areas [35], with the majority in informal spaces with limited services [36]. All of these factors contribute to the vulnerability of low-income groups.

The study starts from the birth of the agglomeration and proceeds with an analysis of its demographic growth, and then the evolution of the built-up areas. Two case studies on a finer scale are presented in order to better understand the socioeconomic conditions of urban areas located in lower areas of the city. These sites were selected based on the identification of sensitive areas affected by flooding during the 2018 rainy season by Service Autonome de Maintenance de la Ville d'Antananarivo (SAMVA). They are among the black spots of the city, since they are flooded every year. They have a similar urban dynamic and socioeconomic situation but differ in terms of the motivation of the people living there. This makes the comparison relevant. The case study analysis is followed by a discussion of results, conclusions, and limitations of the research.

2. Study Area: Agglomeration of Antananarivo

The agglomeration of Antananarivo, also called Greater Antananarivo, is located on the central Malagasy highlands (Figure 1). It covers an area of 76,800 ha and in 2018 had about 2.9 million inhabitants according to INSTAT. On the administrative and institutional level, it brings together the Urban Community of Antananarivo (CUA) composed of six boroughs forming the city of Antananarivo and 37 peripheral municipalities. The whole is located in the Regions of Analamanga and Itasy and is subdivided into 571 neighborhoods called Fokontany.

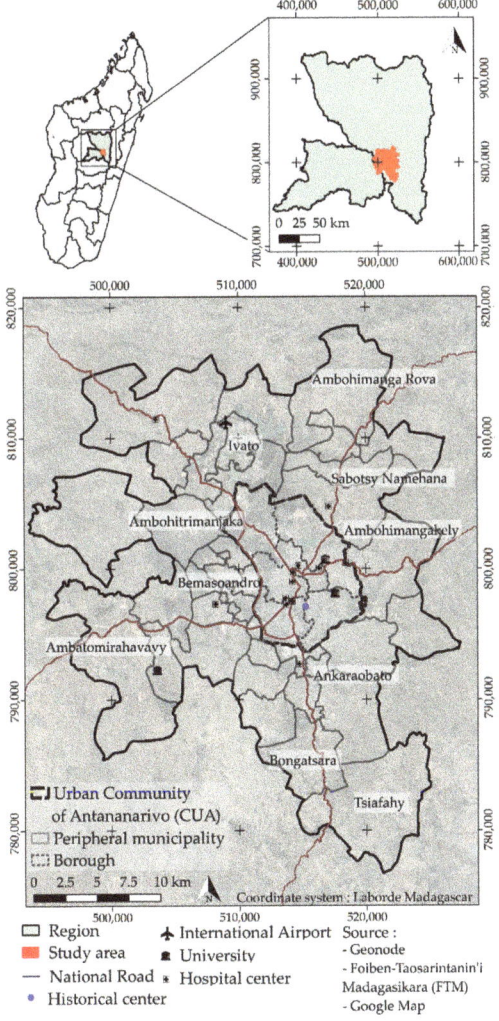

Figure 1. Location of agglomeration of Antananarivo.

Located at an altitude between 1200 and 1500 m above sea level, Antananarivo is characterized by a wide variety of landforms. It is made up of a set of elevated areas with steep slopes to the south, lower areas to the east and center, and a vast alluvial plain in the north and west (Figures 2 and 3).

The plain is drained by the Ikopa and its tributaries (Figures 2 and 3), flowing mainly from the south, southeast, and east to the northwest [37]. Upstream, the flow is more fluid, because the rivers face areas with steep slopes. The river slows down and generates water retention in the lower parts upon its arrival in the plain. This is due to the slight slope of about 0.25% [38] as well as the confluence of the rivers. The topography of the site and its hydrographic network make it very vulnerable to flooding. Almost a third of the urban area is occupied by flood-prone areas (Figure 3).

These flood-prone zones (Figure 3) were produced from a combination of the topographic wetness index (TWI), a soil moisture index, and the stream power index (SPI), an index characterizing the intensity of surface runoff. They are extracted from calculations carried out based on a digital Shuttle Radar Topography Mission (SRTM) model with geographic information system (GIS) software. These two indices are important parameters in flood sensitivity analysis [39].

After the last confluence in the northwest, the Ikopa flows to a single point [34], characterized by a succession of rock outcrops that reduces the water evacuation capacity and generates the formation of alluvial deposits at the level of the plain [40]. In the CUA, the plain forms a polder surrounded by dikes that protect it from overflowing rivers. However, as the river levels are often higher than its level during the rainy season, it is very sensitive to flooding [41].

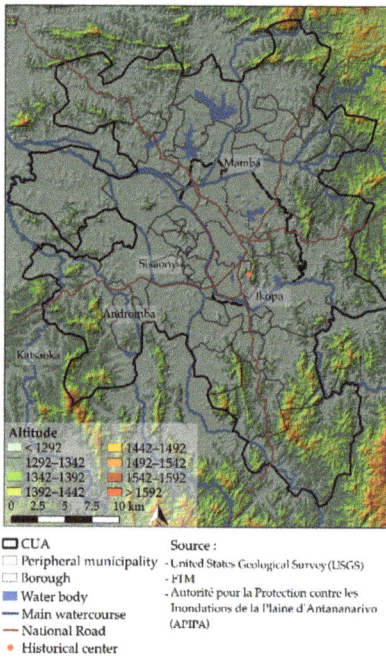

Figure 2. Relief of study area.

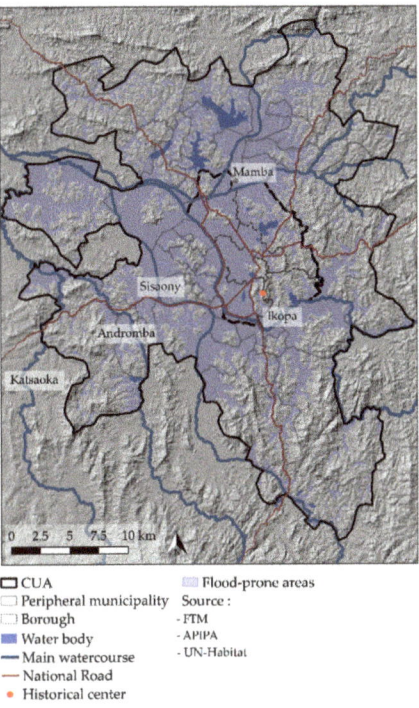

☐ CUA ▨ Flood-prone areas
☐ Peripheral municipality Source :
☐ Borough - FTM
■ Water body - APIPA
— Main watercourse - UN-Habitat
— National Road
• Historical center

Figure 3. Flood-prone areas.

The first hydraulic infrastructures date from the royal period in the 17th and 18th centuries, consisting of river embankments and canals constructed to protect the plain and ensure the evacuation of irrigation water [42]. Currently, the drainage system is denser and more complex. It is structured along three main channels with multiple functions: drainage of rainwater, wastewater, and sewage, as well as irrigation of the agricultural plain. These are the Andriantany, C3, and GR (Génie Rural) channels (Figures 4 and 5). Primary canals, open drains, and buried pipes of various dimensions are connected to these three main channels. Pumping stations and retention basins have been added to this to ensure operation.

At the CUA level, the Andriantany and C3 channels are the main drainage channels. The upstream Andriantany runs through the western part of the city to the pumping station to the northwest and collects water from the hills and eastern plain. It collects rainwater, but also wastewater, from some Fokontany [37]. Downstream, it takes up water from the pumping station and drains to its point of confluence with the Ikopa [43]. As for channel C3, it collects water from the southern plain and agricultural drainage flows as well as excess flows from Andriantany [41]. The GR channel irrigates the plain [44] and acts as a drain during the rainy season [45].

Despite the existence of this drainage system, water tends to accumulate in the plain. At the pumping station, the flow remains paltry compared to the total flow to be drained [46]. Moreover, the increased intensity and volume of runoff due to soil sealing accelerates the degradation of the drainage system [47]. These problems cause flooding during the rainy season.

The peripheral municipalities are not connected to the main drainage network. In these municipalities, water is channeled through sanitation devices along the road network that evacuate rainwater, concrete or earth canals, and drainage ditches leading into the natural environment. In some communes, a collective sanitation system does not even exist [34].

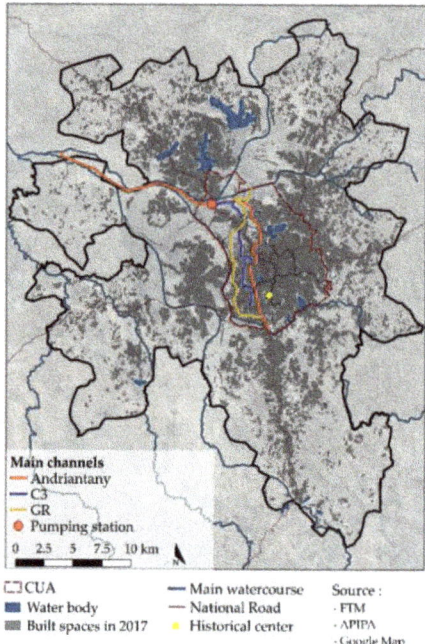

Figure 4. Agglomeration's main drainage system.

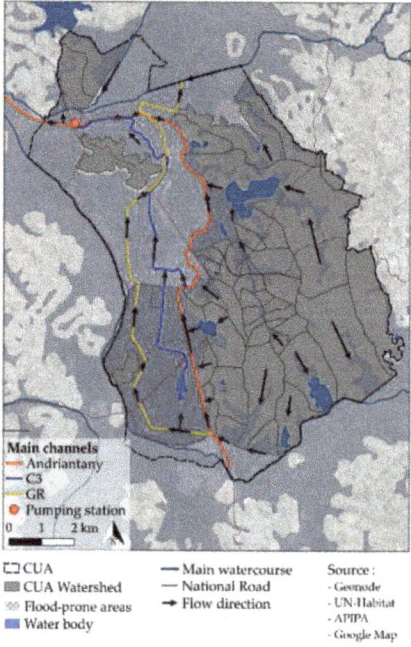

Figure 5. Urban Community of Antananarivo (CUA) watershed.

3. Materials and Methods

3.1. Collected Data

The demographic data used for this work were collected from a general population census based on administrative divisions provided by INSTAT conducted in 1993. For 2018, the figures are based on estimates, since the current general census is not yet official.

Built spaces correspond to structures that host housing, service, industrial, and economic activities. Those represented in this study were digitized based on historical maps and aerial photos. The first base map are maps of Antananarivo from 1953 and 1975. They were provided by Foiben-Taosarintanin'i Madagasikara (FTM), a public establishment in charge of cartography and geographic information in Madagascar. It covers the Greater Antananarivo area except for a few communes to the north and west of the agglomeration. The map representing the areas built in 2006 and 2017 was developed through vectorization of aerial images provided by Google Earth.

The flood-prone zones used in this study are those described in Figure 3 and explained in Section 2. These data were provided by UN-Habitat Madagascar.

Data related to the drainage system came from two organizations that specialize in sanitation and flooding in Madagascar, Autorité pour la Protection contre les Inondations de la Plaine d'Antananarivo (APIPA) and SAMVA.

Table 1 shows the details of these data.

Table 1. Data used in the analysis. INSTAT, Institut National de la Statistique de Madagascar; FTM, Foiben-Taosarintanin'i Madagasikara; APIPA, Autorité pour la Protection contre les Inondations de la Plaine d'Antananarivo; SAMVA, Service Autonome de Maintenance de la Ville d'Antananarivo.

Name of the Data	Spatial Extent	Scale	Date	Source
Demographic data	Agglomeration	-	1993 and 2018	INSTAT
Topographic map	CUA and 32 peripheral municipalities	1/100,000 1/50,000	1953 1975	FTM
Aerial images	Agglomeration	-	2006 and 2017	Google Earth
Flood-prone areas	Agglomeration	-	2012	UN-Habitat Madagascar
Drainage network: main channels	Agglomeration	-	2017	APIPA
Drainage network: secondary channels	CUA	-	2017	SAMVA

In order to cross-check all of the data, cartographic work was carried out. Analysis, modeling, and display of results was done using QGIS geographic information system software.

3.2. Method

The number of inhabitants is a first measure of the degree of exposure to flooding. Demographic data were first mapped at the level of the peripheral municipalities and boroughs of the CUA. This map was used to identify the annual demographic evolution between 1993 and 2018 and the population in 2018. Demographic data were then mapped at the level of the Fokontany in order to obtain the residential population density in 2018 (Scheme 1). Residential population density is defined here as the ratio between the number of inhabitants and the area of built spaces.

Considering that in 2018 the average residential population density of the Fokontany of the agglomeration was estimated at 250 inhab/built ha, the following thresholds were established:

- Fewer than 25 inhab/built ha: very low density
- 25 to 150 inhab/built ha: low density
- 150 to 350 inhab/built ha: moderate density
- 350 to 600 inhab/built ha: densely populated
- 600 to 1200 inhab/built ha: very densely populated
- More than 1200 inhab/built ha: very high density

The density map highlights the overall geographic distribution of the population throughout the agglomeration. Crossed with the area of flood-prone zones of each Fokontany in 2018, it allows estimation of percentage of the population living in flood-prone areas (Scheme 1).

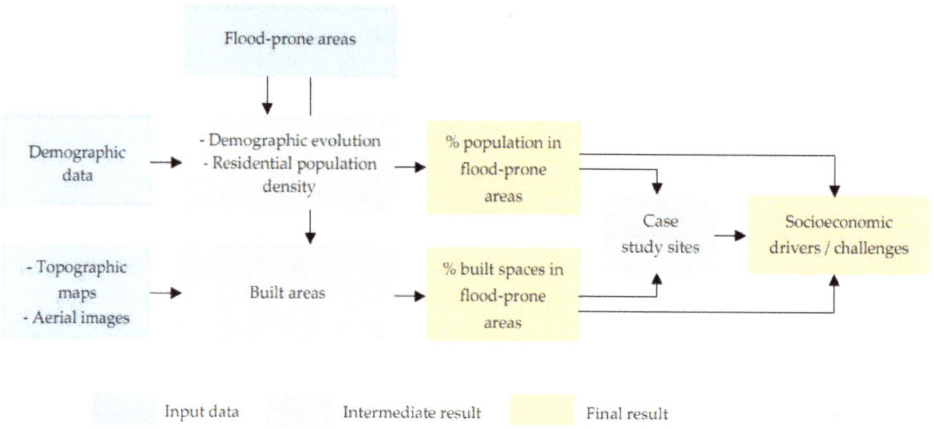

Scheme 1. Overview of method adopted in the study.

Historic topographic maps and recent aerial data were used to map the extent of urbanization in 1953, 1975, 2006, and 2017. This makes it possible to understand the actual structuring of urbanization and the evaluation of the rate of waterproofing of the soil. The urbanization maps were used to determine the progress of construction in flood-prone zones (Scheme 1).

Two groups of Fokontany from the south plain of the CUA were studied in order to understand the factors underlying urbanization in these areas. These Fokontany, Ampefiloha Ambodirano and Ampandrana-Besarety and Besarety, which were selected based on previous analyses, are located in flood-prone areas and witnessed strong demographic growth over the last years. Site visits were made in order to map current land uses and study the local drainage systems. These site visits further allowed us to observe the living conditions in these areas so as to better understand the interplay between flooding, urbanization, and socioeconomic drivers/challenges (Scheme 1).

4. Results

4.1. Demographic Change and Residential Population Density in 2018

Antananarivo experienced considerable demographic growth over the last decades. It was home to more than half of the urban population of the country and around 11.3% of the total population in 2018. Between 1993 and 2018 (25 years), annual growth was estimated at 3.8%.

In 2018, the CUA had approximately 50% of the total population of the agglomeration. However, population growth in the CUA between 1993 and 2018 was rather modest compared to the growth observed in peripheral municipalities (Figure 6). Population growth in peripheral municipalities was mainly driven by migratory flows from the CUA, defined as proximity migration [48], but also from other regions of the country [37].

The demographic density was much higher in the CUA than in the peripheral municipalities in 2018 (Figure 7). It can be observed, however, that the distribution of the population within each commune varies from one Fokontany to another, with some denser nodes outside the CUA.

The highest density values in 2018 are seen in the Fokontany in the center of the CUA and within a radius of 2.5 km (Figure 7). The density peak reached up 2000 inhabitants/built ha, four times the average density of the Fokontany of the CUA and eight times that of the agglomeration. Most of the denser Fokontany were in the western floodplain.

In the peripheral municipalities, densification was led by national roads toward five main axes and grew with an area of expansion around a 10 km radius of downtown CUA (Figure 7). In these communes, the Fokontany were less dense than those of the CUA. However, some of them were in flood-prone areas.

Based on the residential population density and the size of the built-up areas in flood-prone areas in each Fokontany, it is estimated that about 32% of the population of the agglomeration and 43% of the population of the CUA lived in flood-prone areas in 2018.

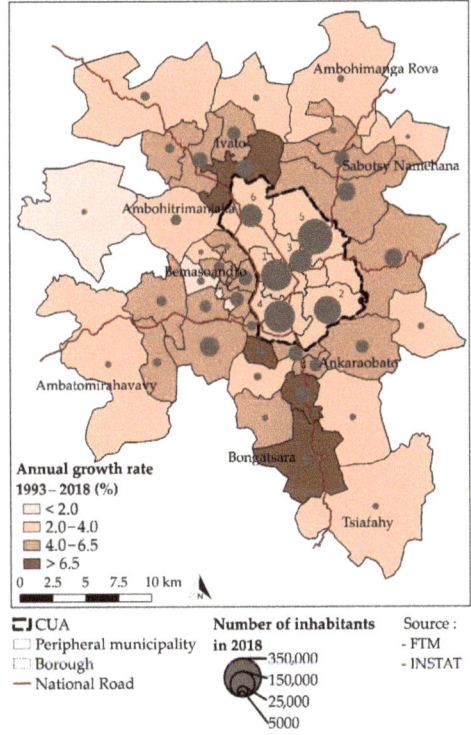

Figure 6. Annual growth rate between 1993 and 2018 and population in 2018 by borough in the CUA and by commune in the outskirts.

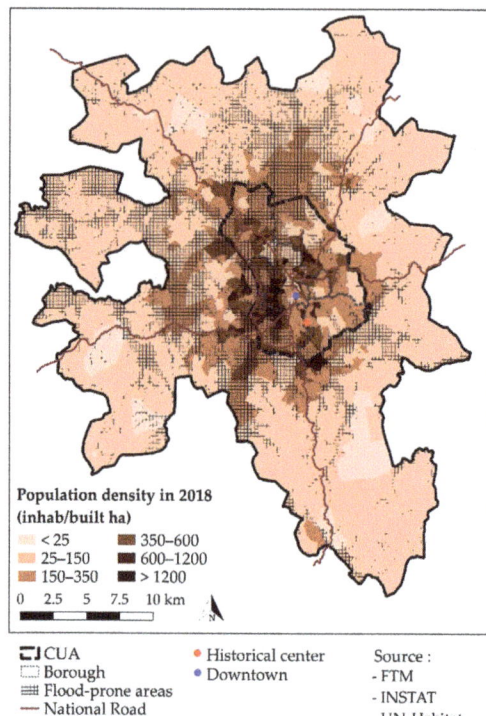

Figure 7. Residential population density in 2018.

4.2. Evolution of Built Spaces

Antananarivo developed in the center of the historic region of Imerina (Figure 8), an ethnic group in the central highlands of Madagascar. The first settlement was the Rova, a royal palace established during the 17th century, on the highest hill in the city [49]. Other constructions were progressively erected around the palace [50]. Some time later, the 30,000 inhabitants of city [51] settled on this first urbanized terrace of the city, called the "upper city" [52]. The extension continued beyond the limit of the hill and spread on the flanks and ridges of neighboring hills, in the north and west, toward the second half of the 19th century and formed the "medium city" [50]. It then gradually developed into the plain, with the installation of small settlement cores in the middle of rice fields [53].

Under the French regime, the extent of this encroachment became more important, and Antananarivo underwent its first major urban transformations. Some 20 hectares were backfilled in order to form the first Fokontany in the "lower town", and other new neighborhoods were created [54,55]. Between 1896 and 1903, 35 km of paved roadways were opened, tunnels were dug, places were created [56], and work on railway lines started [57]. This was only the beginning of the urbanization that would occupy the entire lower area a few years later.

In 1953, the built space mainly occupied elevated areas and covered around 1806 ha, i.e., 2.4% of the area of the present agglomeration (Table 2). In the CUA, urbanization was mainly oriented toward the east, around the historic center (Figure 8). On the other hand, the development of buildings in the lower town of the west continued. 10% of the CUA area was urbanized, and some 2.6% of the area was then covered by built areas located in flood-prone zones (Table 2). Outside the CUA, the development line progressed gradually to the east and northwest. Throughout the agglomeration, spatial development operated through the densification of spaces that were already built and through "fingerprint" urbanization, where most constructions were arranged along the main national axes (Figures 8 and 9). Among all built areas, 22% were in flood-prone zones, covering an area of 399 ha (Table 2).

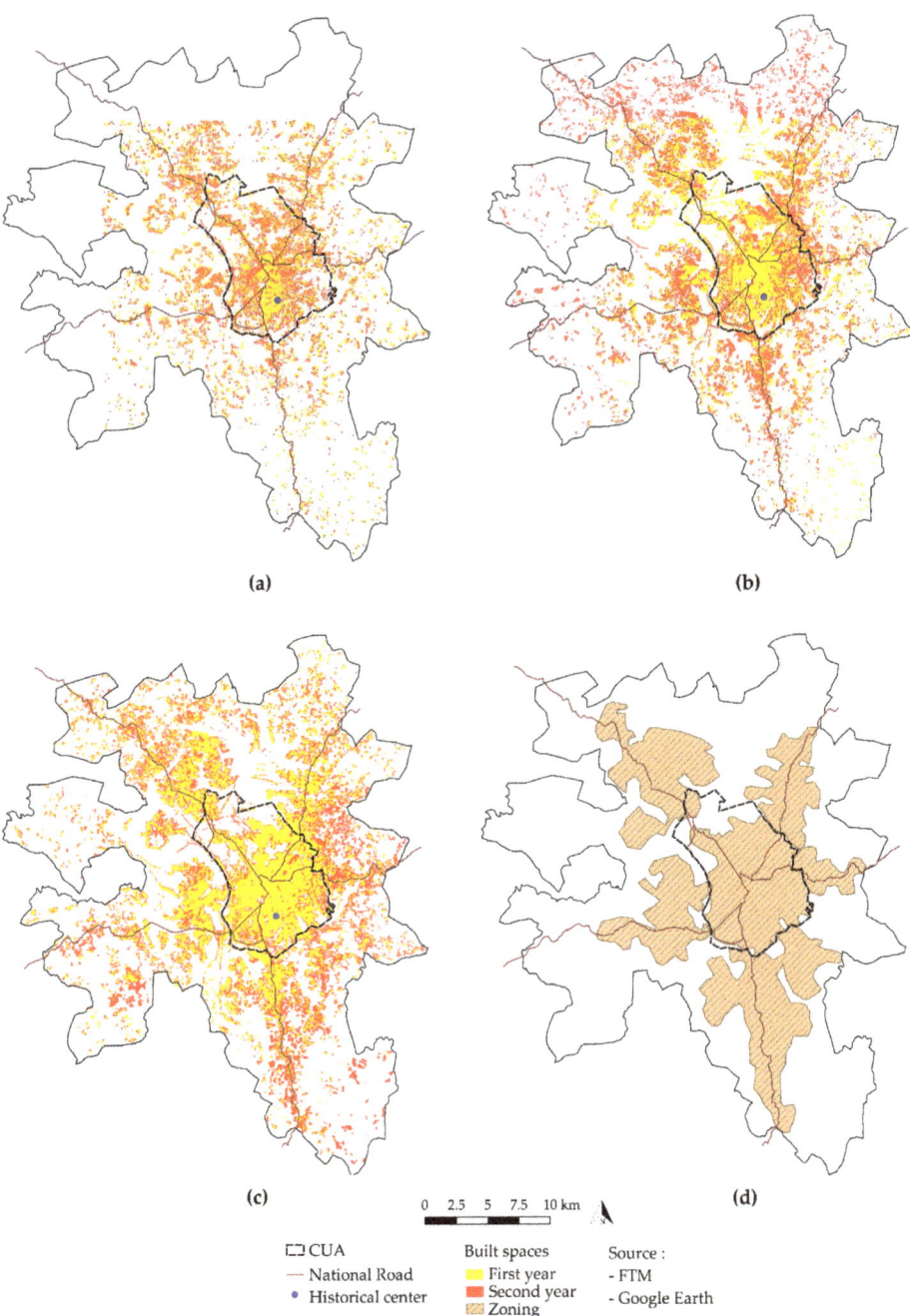

Figure 8. Urbanized cells: (**a**) 1953–1975; (**b**) 1975–2006; (**c**) 2006–2017; (**d**) 2017.

Table 2. Extent and increase of urbanized cells and built areas in flood-prone zones.

Date	Built Areas					Built Areas in Flood-Prone Zones				
	CUA	Agglomeration	Total Increase	Yearly Increase		CUA	Agglomeration	Total Increase	Yearly Increase	Compared to Built Spaces
	(%)	(%)	(ha)	(%)	(%)	(%)	(%)	(ha)	(%)	(%)
1953	10	2.4	1806			2.6	0.5	399		22
1975	25.5	5.4	4143	129	5.9	8.3	1.4	1101	176	8	26.5
2006	48.2	10.5	8048	94	3	13.9	2.7	2038	85	2.7	25
2017	55.3	21.2	16,250	102	9.3	19.8	4.8	3675	80	7.3	23
1953–2017					6.1					6

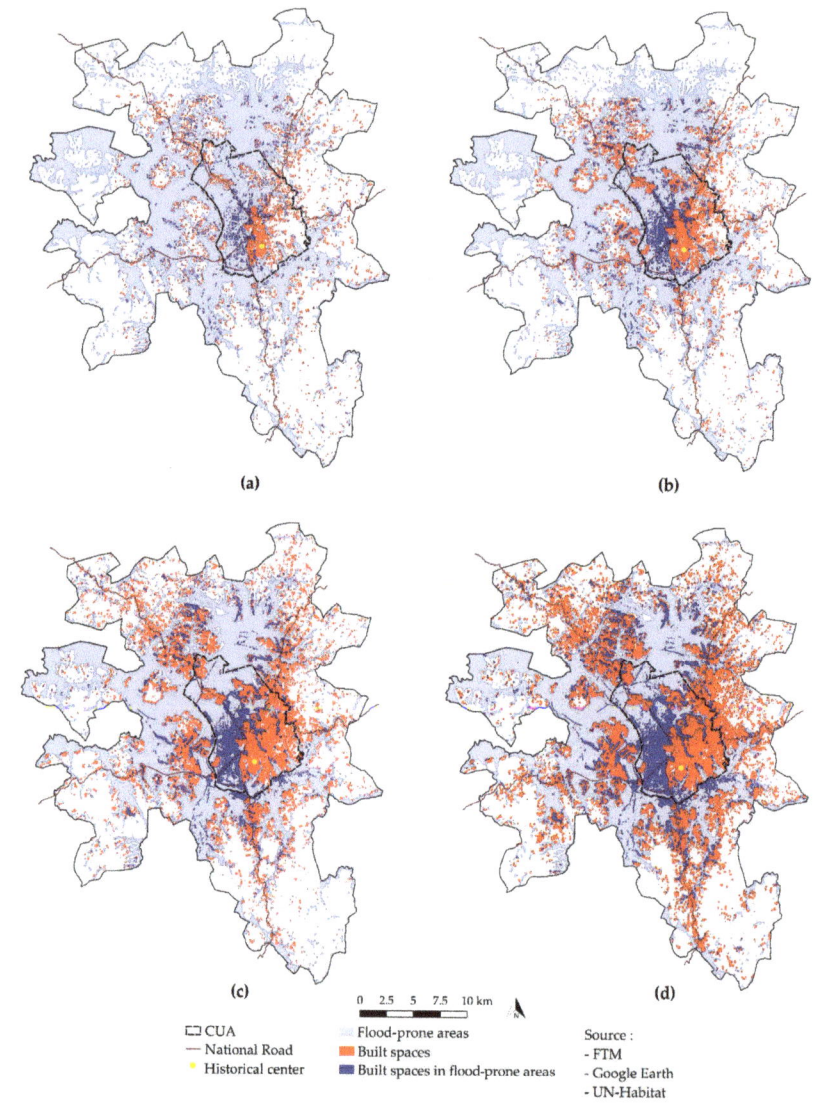

Figure 9. Urbanized cells in flood-prone zones: (a) 1953; (b) 1975; (c) 2006; (d) 2017.

From the 1960s, the period of independence, backfilling continued in the flood plain (Figures 8 and 9). Several new urban areas were built, and this did not protect the city from flooding. Urbanization continued until 1975, with approximately 2330 ha more of built areas than in 1953, and built areas then covered 5.4% of the agglomeration. The increase compared to 1953 is estimated at 129%, an annual increase of 5.9% (Table 2). Inside the CUA, built areas grew considerably and covered 25.5%. These are divided into two distinct parts: to the east toward the elevated areas and to the west in the flood plain (Figure 9). Built areas in flood-prone zones then represented 8.3% of the CUA, a sharp increase compared to the situation in 1953. In peripheral municipalities, urbanization mainly occurred through the filling of voids and the densification of existing areas. In addition to this urban expansion at the agglomeration level, the share of built areas located in flood-prone zones increased very rapidly. Compared to 1953, built areas located in flood-prone zones increased by 176%, an annual increase of 8%, covering 26.5% of the entire built area at the agglomeration level (Table 2).

From 1975 to 2006, urban expansion progressively shifted toward peripheral municipalities, especially in the neighboring municipalities of the CUA (Figure 8). It was guided by national roads and showed five centers of urban growth. In the CUA, urban consolidation continued and densified the area to the east and northwest. Built areas then occupied 48.2% of the CUA. In the whole agglomeration, between 1975 and 2006, the built areas increased by 94%, corresponding to an annual increase rate of 3%. The built areas then occupied 10.5% of the agglomeration, i.e., 8048 ha. It can be seen from Table 2 that 25% of built areas were then located in flood-prone zones. This is equivalent to an 85% increase compared to the situation of 1975, i.e., a 2.7% yearly increase over the period 1975–2006 (Table 2).

Between 2006 and 2017, urban expansion further intensified, especially in peripheral areas. Constructions was dispersed in areas far from the CUA. Approaching the CUA, the urban fabric became denser, especially in the northwest and the south (Figure 8). In the CUA, despite high density, urbanization continued to progress, to a large extent at the expense of rice fields in the lower town (Figure 8), by filling interstices to the west and rising slightly toward empty spaces to the north (Figures 8 and 9). A total of 55.3% of the CUA was occupied by built areas and 19.8% was occupied by built areas located in flood-prone zones. At the agglomeration level, built areas covered 16,250 ha, or 21.2% of the territory. It doubled over the years 2006–2017. The increase of built areas since 2006 was estimated at 102%, i.e., an annual increase of 9.3%. Constructions kept settling in the flood plain, and 23% of the total built areas was located in flood-prone zones in 2017. With an increase rate of 80%, or 7.3% yearly, built areas located in flood-prone zones occupied 4.8% of the agglomeration (Table 2).

Between 1953 and 2017, the annual rate of increase of built areas in the agglomeration was 6.1%. In flood-prone zones, it was estimated at 6% (Table 2). Urban expansion in flood-prone areas developed in parallel with that in other areas of the agglomeration. Practically, it means that building in flood-prone zones in Antananarivo was driven by the general expansion of the city, which was intense.

4.3. Case Studies

4.3.1. Case Study 1: Fokontany of Ambodirano Ampefiloha

The lower neighborhoods of the west, including the Fokontany of Ambodirano Ampefiloha, are the Fokontany located in the lower town on the left bank of the Andriantany canal (Figure 10). They constitute the extension of backfilled spaces in the plain of Antananarivo during the colonial period. Urbanization in these places accelerated following major subdivision operations in the 1970s. Apart from these constructions and a few service buildings, most houses are made of recycled materials such as wood, plastic, or brick and are in very poor condition [43] and underserved. The area gathers populations with low income, which prevents them from accessing other housing [54]. Homes are cramped and overcrowding prevails. They are exposed to flooding due to river floods and even overflows of drainage canals, and to health risks. These are not negligible due to the almost permanent accumulation of water during the rainy season.

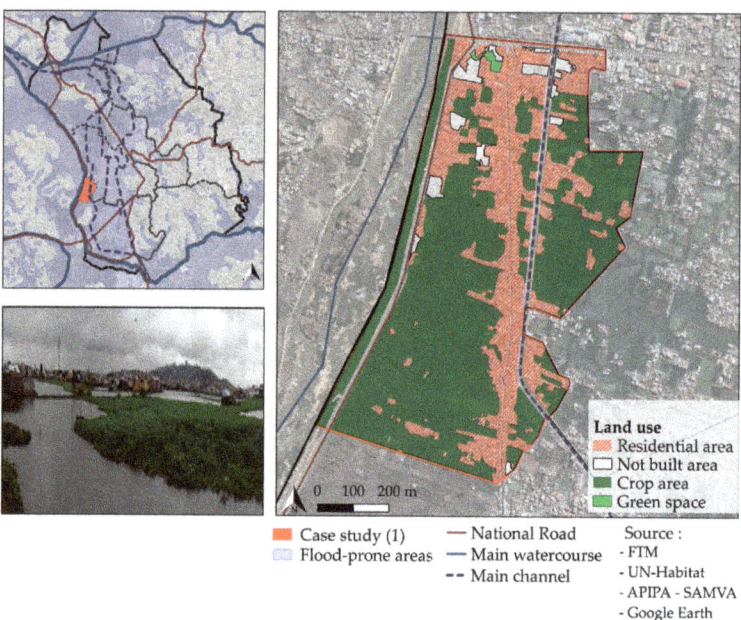

Figure 10. Map of Fokontany of Ambodirano Ampefiloha in CUA.

Ambodirano Ampefiloha covers an area of 63 ha. It is crossed in the east by the GR channel, the irrigation channel of the agricultural plain, and bordered by the Ikopa in the west. It has a flat topography with areas that form basins accumulating large volumes of water.

Between 1993 and 2018, the population almost tripled. With the growing population, this Fokontany is densely populated, with a residential density estimated at 484 inhabitants/built ha. The dynamics of family migration based on family support to overcome the difficulties encountered in the rural world is a main source of this considerable densification [43]. The residential area occupies 30% of the total area and rice fields 60%. The Fokontany does not have a specific drainage network (SAMVA).

This large urban space should contribute to the storage of water during rainy periods [58]. Nevertheless, it is increasingly invaded by constructions.

4.3.2. Case Study 2: Fokontany of Ampandrana-Besarety and Besarety

The Fokontany of Ampandrana-Besarety and Besarety are classified among the working-class neighborhoods at the foot of the upper town (Figure 11). They are located at altitudes between 1251 and 1252 m and extend over 22 ha. As in most of the Fokontany in the lower areas, backfilling allowed a progressive urbanization of the zone [59]. Floods are mainly linked to the overflow of drainage channels.

With a residential population density of 548 inhab/built ha, well above the CUA average, these Fokontany are categorized as densely populated. Population growth was 2.9% per year between 1993 and 2018. This upsurge was due to the high birth rate and the arrival of new migrants. The proximity to the city center and industrial zones explains why new inhabitants have come to settle there.

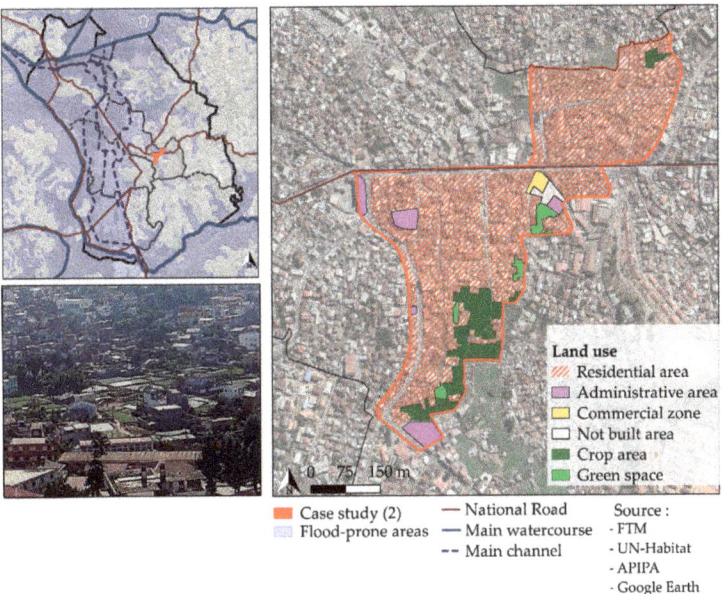

Figure 11. Map of Fokontany of Ampandrana-Besarety and Besarety in CUA.

Mainly residential activities and constructions are developing, to the detriment of rice fields. Two categories of settlements can be identified here: traditional or modern concrete constructions and small constructions made of sheet metal or wood. Currently, dwellings cover 75% of the total area, and 7% is occupied by cultivated areas. However, the whole area has a waterproofing rate of more than 80%, which generates large runoff.

The drainage system works differently on both sides of the area (SAMVA). It is operated through underground networks to the west and provided by gutters in the east. The whole system is subsequently taken up by primary channels. Despite the existence of this system, the primary network remains constrained by the accumulation of water with an important flow upstream of the site and by the weak slopes of the plain. Furthermore, the sections of these channels are heterogeneous. In certain sections, the load is much greater. In addition, the flow of water is hampered by the congestion of gutters and the obstruction of manholes by deposits of sand or waste [43]. Some installed structures also make it more difficult for water to flow during rainy events [60]. All of this, together with the proximity of the constructions to the drainage network and even their encroachment on the network, makes these Fokontany areas with recurrent floods.

5. Discussion

In SSA, several urban areas have experienced high population growth in recent decades. Gardi has shown that 12 of the top 30 fastest-growing urban agglomerations in the world are in SSA [17]. Antananarivo ranks 20th in the world and 9th in SSA in this ranking, right next to the major cities of Nairobi, with an annual rate of 3.87%, and Kinshasa, with 3.89% [17]. This growth is due to the high rate of natural increase and the rural exodus to and near urban areas [36]. In SSA, internal growth has been a determining factor for many years [16,61,62]. Nevertheless, migration is also part of the driving force behind population growth and urban sprawl [63–65]. For sociocultural, economic, political, and environmental reasons, whether it is a choice or a necessity, this migration seems to be a way to allow better living conditions for rural migrants [63,66]. In South Africa, due to inadequate social services and the lack of employment in rural areas, people migrate to urban areas [66,67]. In the Democratic Republic of Congo, it is because of conflicts and insecurity [65,68]. In Burkina Faso and Kenya, it is due to climatic disturbances [69]. The expansion of informal activities [66] and the concentration of services and facilities in cities also easily attract the rural population [65,70]. In Madagascar, social and land insecurity in the countryside [45] as well as economic difficulties [71] due to declining agricultural productivity [34] are pushing rural populations to migrate to Antananarivo. Due to the accessibility of services and infrastructure, students, civil servants, and people involved in small trade migrate and increase the size of the city's population [53].

This demographic growth creates unease for urban centers, since the supply of housing and spaces to be built does not meet the growing demand [63,72–74]. The fragility of urban governance, manifested in the lack of support and prioritization of urban infrastructure and land-use management initiatives by governments, is one of the reasons for this [75]. Parnell, Pieterse, and Watson also refer to a lack of good planning [20] due to poorly conceived planning laws and standards for construction as well as insufficient funding [16]. As a result, many informal settlements are developed [18]. The proliferation of these informal settlements is also related to increasing urban poverty and leads to the involvement of many poor people in the informal economy [65,76]. On the other hand, it is produced by the proximity of informal employment, which pushes migrants and poor households to settle nearby [77,78]. It is also accompanied by the development of new housing areas in flood-prone areas [18,19,79]. These settlements are generally precarious, with poor infrastructure, and are heavily impacted by flooding [18,70,80]. For Antananarivo, given the weakness of urban planning and the absence of housing policies, the demand greatly exceeds the supply that the city can offer [81]. This has led to illegal installations and constructions in the west of the city, in the flood plain [82]. In order to overcome this deficit, public authorities indirectly approved and anticipated the extension of urbanization in the plain [83].

In SSA, most of the urban population live in informal settlements in areas at risk [23], such as floodplains, swamps, and riverbanks [84,85]. This population group mainly consists of the urban poor [84]. They are often excluded from the land market due to unaffordable prices and imposed standards and regulations that they are unable to follow, and that force them to occupy these dangerous lands [11,21,86]. In addition to this group are refugees and persons who are displaced due to forced displacement in cities who settle in these areas for various reasons [87–90]. These situations and the increased population density in these areas make them more vulnerable to flooding [84,89,91].

This set of processes is linked to the history of urbanization in SSA. In the region, the pre-colonial period is characterized by a low level of urban development [76] that intensified during the period of colonization [16,65,76]. The occupation of colonial cities by settlers [76,79] favored the development of informal settlements formed by the indigenous population, who were excluded from planning [92]. In Nairobi, this led to the construction of several illegal settlements by homeless Africans who were only allowed to be in the city for work [92]. It also promoted the development of habitats in flood-prone areas, as in the case of Antananarivo. In the 1930s and 1960s, development of the colonial city led to the displacement of several population groups, who took refuge in low areas [93]. In addition, the overcrowded hills of the middle city and the conveniences of the plain attracted inhabitants of the upper city to the lower city [53]. In the 1950s and 1960s, urbanization was mainly fueled by the development of industrial facilities [43], part of the orientation of the 1954 colonial-era urban plan [83]. These works changed the hydraulic regime of the plain and led to densification of the flood plains [59]. Flood-prone zones also became more favorable for speculators and real estate developers. Actually, building is less onerous and investments more profitable in low lands, given the more suitable topography [83]. By contrast, the extension of existing habitats and the widening of service roads were more expensive in the eastern part, given the steep slopes (up to 20%) [40,94]. There are also landslide risks in this part of the city [95].

In the years following independence, around 1960 and 1970, many cities in SSA experienced rapid population growth [65,79,96]. This growth can be linked to policies adopted after independence, which were related to the deployment of jobs, the establishment of several industries in city centers [65,97], and investment in public works [98]. The average annual population growth rate in urban areas was approximately 5% [99]. However, the cities inherited from the colonial era were not designed to accommodate such a massive population [20,76]. This reinforced the proliferation of informal settlements [76] that are more exposed to flooding [74]. In Accra, Ghana, the informal development of some communities in watersheds around rivers and lagoons increased their exposure to flooding [100]. In addition, the economic crisis that hit Africa in the 1970s fueled such difficulties throughout the region [101]. There has been a decline in investment in urban infrastructure and housing [76]. For Antananarivo, the 1970s were characterized by the completion of development and subdivision work to replace the thousands of homes destroyed following the devastating 1959 floods [83,102]. Other districts were then created in the lower town to accommodate the affected populations [94]. As these social housing units were not affordable for vulnerable populations, the construction of precarious housing proliferated. Antananarivo was also plunged into a lasting multifaceted crisis [56]. The construction sector was strongly affected. Materials were more expensive and scarcer. This led to the proliferation of informal habitats in flood-prone zones [43]. This was favored by the absence or inadequacy of urban land management tools [103]. It is difficult to access land due to expensive administrative procedures and a lack of updated information about the legal status of the land [45].

From the 1990s, urban development accelerated, especially in the peripheral areas around cities [62]. This is partly due to the decline in land values in these areas [104]. However, these habitats are often built outside of planning and regulations [65,104,105], and are places where various risks, including flooding, are prevalent [23]. For Antananarivo, the emergence of several peri-urban cores outside the CUA was supported by the establishment of the city's 2004 urban plan, which proposed unclogging the city center [37]. The migratory flow consequently became more important toward peripheral municipalities. This is further related to the limited accommodation capacity of CUA houses linked to their architecture [106] and the lack of available land for development in residential areas [107]. Peripheral areas were more attractive for residential installations in terms of both availability of building land and cost of living [108]. The establishment of industrial buildings [46] and the proliferation of infrastructure projects led to the acceleration of urbanization, which is increasingly taking place in flood-prone areas where land is cheaper and rents moderate [82,106]. The proximity of these settlements to industrial areas attracts the population, as shown in the case studies. These factors were favored by the absence of urban planning for a long period (from 1968 to 2004) [83].

This growing urbanization in SSA is also accompanied by a lack of infrastructure, including drainage systems, which makes cities vulnerable [19,79,91]. In Antananarivo, the case studies reveal this. The proximity of constructions to drainage canals and informal encroachments on these canals progressively reduce their capacity [60]. As the load on the existing network increases, this leads to greater susceptibility to the effects of further flooding [74].

Scheme 2 synthesizes the co-evolution of urbanization and vulnerability of poorly managed urban environments. The demographic pressure leads to urban sprawl around major cities (1). Due to the lack of housing and appropriate urban management (2), the expansion of the city occurs through informal settlements (3). It is more the case that incoming populations, especially from rural areas, are usually associated with low economic resources (4). Flood-prone areas are associated with the development of informal settlements (5): land is cheaper, and constructions are not authorized by planning documents. Furthermore, flood-prone areas located near canals and rivers are well adapted to maintaining subsidence urban agriculture, especially for inhabitants coming from rural areas. The construction of precarious settlements in flood-prone areas (6) by low-income groups (7) obviously exacerbates the vulnerability of these groups and their habitat; even more as the drainage system in these areas is often insufficient (8), which can partly be explained by the fast urban growth witnessed by SSA cities (9) and the lack of adequate integration of drainage in urban planning policies (10). The lack of drainage combined with urban sprawl and soil-sealing further contribute to increasing floods at the agglomeration level (11).

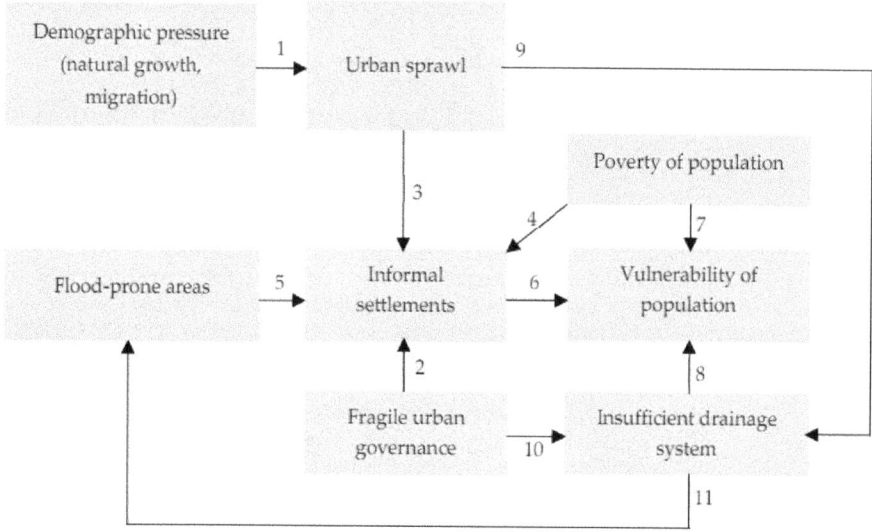

Scheme 2. Flood vulnerability factors, a co-evolutionary perspective.

6. Conclusions

Flooding is an occurrence that plagues most countries in the world, particularly in urban centers in SSA. Over time, poorly planned urbanization, combined with several other factors, expose people and buildings to flooding and increase their vulnerability. The agglomeration of Antananarivo is a clear example. Like many African urban agglomerations, it faces strong demographic pressure due to the high rate of natural growth and population migration toward urban areas. The city progressively expanded in the lower zone without much control. Our analysis shows that around 32% of the population of the agglomeration lived in flood-prone zones in 2018. The annual growth of built-up areas in flood-prone zones between 1953 and 2017 is estimated at 6%. In 2017, 23% of the buildings of the agglomeration, i.e., almost one out of four buildings, were in flood-prone zones.

The dynamics of urbanization inherited from the colonial era led over time to the proliferation of informal settlements, which are more exposed to flooding. This led to a modification of the hydraulic system of the city and a high degree of vulnerability in the flood plain. Given the inadequacy of the drainage infrastructure, the agglomeration is suffering from increasingly harsh flood events, especially during the rainy period. This is a trend known and present in the literature on urban agglomerations in SSA.

Faced with the rapid growth of the population, which leads to high demand in terms of constructions, urban planning and management are essential to avoid installations in flood-prone areas. This concerns dwellings as well as other structures that may attract inhabitants. The inadequacy or even the absence of planning, most of the time accompanied by a lack of provision of adequate services and equipment, leads to concentrated precariousness in flood-prone zones, where land is obviously cheaper. Integrating flood risk management in spatial planning policies is essential to curb this phenomenon. It should be combined with proper implementation of these policies, targeting both newly developed and existing urban areas. Since many people live in flood-prone areas, it would be more relevant to work on resilience to reduce vulnerability than to relocate this large volume of people.

7. Limitations of the Research

In the study of flooding in urban areas, it is advisable to consider the different parameters mentioned above to understand the issue and be able to propose solutions to reduce vulnerability. In the framework of our research, we focused on the product of urbanization and its combination with other parameters that promote vulnerability. We did not consider here some factors that may contribute to increase urban flooding, such as climate change. The flood-prone areas considered in our study remain constant over time, because they are based on the calculation of topographic indices (TWI and SPI) relative to the topography. The study does not consider either the amount of precipitation or the flood history. Finally, in our paper, we considered urbanization relative to the whole of the built spaces. It would be relevant to consider built spaces that are only residential areas and integrate land use and land use change.

Author Contributions: Conceptualization and methodology: F.N.R. and J.T.; software, validation, and formal analysis: F.N.R.; investigation and resources: F.N.R. and J.T.; data curation: F.N.R.; writing—original draft preparation: F.N.R. and J.T.; writing—review and editing: F.N.R. and J.T.; visualization: F.N.R. and J.T.; supervision: J.T.; project administration: F.N.R. All authors have read and agreed to the published version of the manuscript.

Funding: This research received no external funding.

Institutional Review Board Statement: Not applicable.

Informed Consent Statement: Not applicable.

Data Availability Statement: Data is contained within the article. They are also available on request from the corresponding author.

Acknowledgments: We are grateful to M. Faly Rabemanantsoa, M. Jaotiana Rasolomamonjy, and M. Jean Ruffin Ramiaramanana for their help in carrying out this work and providing relevant data.

Conflicts of Interest: The authors declare no conflict of interest.

References

1. White, I. *Water and the City: Risk, Resilience and Planning for a Sustainable Future*; Routledge: Abingdon, UK, 2013.
2. Guha-Sapir, D.; Hargitt, D.; Hoyois, P. *Thirty Years of Natural Disasters 1974–2003: The Numbers*; Presses Univ. de Louvain: Ottignies-Louvain-la-Neuve, Belgium, 2004.
3. CRED. *Natural Disasters 2018*; CRED: Brussels, Belgium, 2019.
4. Kundzewicz, Z.W.; Kanae, S.; Seneviratne, S.I.; Handmer, J.; Nicholls, N.; Peduzzi, P.; Mechler, R.; Bouwer, L.M.; Arnell, N.; Mach, K. Flood Risk and Climate Change: Global and Regional Perspectives. *Hydrol. Sci. J.* **2014**, *59*, 1–28. [CrossRef]
5. Hua, P.; Yang, W.; Qi, X.; Jiang, S.; Xie, J.; Gu, X.; Li, H.; Zhang, J.; Krebs, P. Evaluating the Effect of Urban Flooding Reduction Strategies in Response to Design Rainfall and Low Impact Development. *J. Clean. Prod.* **2020**, *242*, 118515. [CrossRef]
6. Bates, B.C.; Kundzewicz, Z.W.; Wu, S.; Palutikof, J.P. Le Changement Climatique et l'eau. *Doc. Tech. Publ. Par Groupe D'experts Intergouv. Sur L'évolution Clim. Secrétariat GIEC Genève Éd* **2008**, 236.
7. Kadomura, H. Climate Anomalies and Extreme Events in Africa in 2003, Including Heavy Rains and Floods That Occurred during Northern Hemisphere Summer. *Afr. Study Monogr.* **2005**. [CrossRef]
8. Marchi, L.; Borga, M.; Preciso, E.; Gaume, E. Characterisation of Selected Extreme Flash Floods in Europe and Implications for Flood Risk Management. *J. Hydrol.* **2010**, *394*, 118–133. [CrossRef]
9. Herring, S.C.; Hoerling, M.P.; Peterson, T.C.; Stott, P.A. Explaining Extreme Events of 2013 from a Climate Perspective. *Bull. Am. Meteorol. Soc.* **2014**, *95*, S1–S104. [CrossRef]
10. Ahiablame, L.; Shakya, R. Modeling Flood Reduction Effects of Low Impact Development at a Watershed Scale. *J. Environ. Manag.* **2016**, *171*, 81–91. [CrossRef]
11. Jha, A.K.; Bloch, R.; Lamond, J. *Cities and Flooding: A Guide to Integrated Urban Flood Risk Management for the 21st Century*; The World Bank: Washington, DC, USA, 2012.

12. Li, G.-F.; Xiang, X.-Y.; Tong, Y.-Y.; Wang, H.-M. Impact Assessment of Urbanization on Flood Risk in the Yangtze River Delta. *Stoch. Environ. Res. Risk Assess.* **2013**, *27*, 1683–1693. [CrossRef]
13. Chocat, B. Le Rôle Possible de l'urbanisation Dans l'aggravation Du Risque d'inondation: L'exemple de l'Yseron à Lyon/The Potential Role of Urbanization in Increasing the Risk of Flooding: The Example of the Yzeron in Lyon. *Géocarrefour* **1997**, *72*, 273–280. [CrossRef]
14. Douglas, I. Flooding in African Cities, Scales of Causes, Teleconnections, Risks, Vulnerability and Impacts. *Int. J. Disaster Risk Reduct.* **2017**, *26*, 34–42. [CrossRef]
15. Diakakis, M.; Pallikarakis, A.; Katsetsiadou, K. Using a Spatio-Temporal Gis Database to Monitor the Spatial Evolution of Urban Flooding Phenomena. The Case of Athens Metropolitan Area in Greece. *ISPRS Int. J. Geo-Inf.* **2014**, *3*, 96–109. [CrossRef]
16. Spence, M.; Annez, P.C.; Buckley, R.M. *Urbanization and Growth*; World Bank: Washington, DC, USA, 2008.
17. Gardi, C. *Urban Expansion, Land Cover and Soil Ecosystem Services*; Taylor & Francis: Abingdon, UK, 2017.
18. Pauleit, S.; Coly, A.; Fohlmeister, S.; Gasparini, P.; Jorgensen, G.; Kabisch, S.; Kombe, W.J.; Lindley, S.; Simonis, I.; Yeshitela, K. *Urban Vulnerability and Climate Change in Africa; Future City*; Springer: Berlin, Germany, 2015; Volume 4.
19. Adelekan, I.O. Vulnerability of Poor Urban Coastal Communities to Flooding in Lagos, Nigeria. *Environ. Urban.* **2010**, *22*, 433–450. [CrossRef]
20. Parnell, S.; Pieterse, E.; Watson, V. Planning for Cities in the Global South: An African Research Agenda for Sustainable Human Settlements. *Prog. Plan.* **2009**, *72*, 233–241.
21. Dodman, D.; Leck, H.; Rusca, M.; Colenbrander, S. African Urbanisation and Urbanism: Implications for Risk Accumulation and Reduction. *Int. J. Disaster Risk Reduct.* **2017**, *26*, 7–15. [CrossRef]
22. Xie, J.; Chen, H.; Liao, Z.; Gu, X.; Zhu, D.; Zhang, J. An Integrated Assessment of Urban Flooding Mitigation Strategies for Robust Decision Making. *Environ. Model. Softw.* **2017**, *95*, 143–155. [CrossRef]
23. Pelling, M.; Wisner, B. *Disaster Risk Reduction: Cases from Urban Africa*; Routledge: Abingdon, UK, 2012.
24. CRED. *The Human Cost of Natural Disasters: A Global Perspective*; CRED: Brussels, Belgium, 2015.
25. Pusch, C.; Bedane, A.W.Y.; Agosti, A.; Carletto, A.L.; Tiwari, A.; Parvez, A.; Dingel, C.C.; Chararnsuk, C.; Wielinga, D.G.; Muraya, F.M. *Striving toward Disaster Resilient Development in Sub-Saharan Africa: Strategic Framework 2016–2020*; The World Bank: Washington, DC, USA, 2016.
26. Tiepolo, M. Flood risk reduction and climate change in large cities south of the Sahara. In *Climate Change Vulnerability in Southern African Cities*; Springer: Berlin, Germany, 2014; pp. 19–36.
27. Cohen, B. Urbanization in Developing Countries: Current Trends, Future Projections, and Key Challenges for Sustainability. *Technol. Soc.* **2006**, *28*, 63–80. [CrossRef]
28. Silva, C.N. *Urban Planning in Sub-Saharan Africa*; Routledge: Abingdon, UK, 2015.
29. Sivapalan, M.; Savenije, H.H.; Blöschl, G. Socio-Hydrology: A New Science of People and Water. *Hydrol Process* **2012**, *26*, 1270–1276. [CrossRef]
30. Sivapalan, M.; Blöschl, G. Time Scale Interactions and the Coevolution of Humans and Water. *Water Resour. Res.* **2015**, *51*, 6988–7022. [CrossRef]
31. DGM. *Le Changement Climatique à Madagascar*; DGM: Antananarivo, Madagascar, 2008.
32. INSTAT. *Troisième Recensement Général de La Population et de l'Habitation (RGPH-3)—Madagascar*; INSTAT: Antananarivo, Madagascar, 2019.
33. Sommer, K.; Guiébo, J.; Vignol, R.; Maréchal, N.; Sublet, M.; Kuria, F. *Madagascar, Profil Urbain National*; ONU-Habitat: Nairobi, Kenya, 2012.
34. ARTELIA. *Elaboration Du Schéma Directeur d'Assainissement Urbain Du Grand Tana*; ARTELIA: Antananarivo, Madagascar, 2014.
35. CPGU & BNCCC. *PPCR Strategic Program for Climate Resilience for Madagascar*; CPGU & BNCCC: Antananarivo, Madagascar, 2017.
36. USAID. *Risques Climatiques Dans Les Zones Urbaines et Voie D'urbanisation*; USAID: Washington, DC, USA, 2018.
37. Raveloarison, T. *Plan D'urbanisme Directeur 2004 Horizon 2015*; Commune Urbaine d'Antananarivo: Antananarivo, Madagascar, 2004.
38. Douessin, R. Géographie Agraire Des Plaines de Tananarive. *Madag. Rev. Géographie* **1974**, 9–156.
39. Bui, D.T.; Pradhan, B.; Nampak, H.; Bui, Q.-T.; Tran, Q.-A.; Nguyen, Q.-P. Hybrid Artificial Intelligence Approach Based on Neural Fuzzy Inference Model and Metaheuristic Optimization for Flood Susceptibilitgy Modeling in a High-Frequency Tropical Cyclone Area Using GIS. *J. Hydrol.* **2016**, *540*, 317–330.
40. M2PATE. *Schéma National d'Aménagement Du Territoire (SNAT) 2015–2025*; M2PATE: Antananarivo, Madagascar, 2015.
41. APIPA. *Le Système de Drainage de La Ville d'Antananarivo: Note Guide*; APIPA: Antananarivo, Madagascar, 2006.
42. Isnard, H. Les Plaines de Tananarive. *Les Cahiers d'Outre-Mer* **1955**, *8*, 5–29. [CrossRef]
43. Labatte, B.; Recouvreur, R.; Estienne, C.; Fernandez, D.; Janssen, J.; Le Cam, Q.; Olivier, F.; Eudora, A.; Olivier, D.; Marty, D.; et al. *Programme Intégré d'Assainissement d'Antananarivo (PIAA)*; BRL Ingénierie: Nimes, France, 2018.
44. GERSAR-BRL. *Manuel de Gestion et d'exploitation Du Nouveau Canal d'irrigation "GR"*; GERSAR-BRL: Antananarivo, Madagascar, 2000.

45. M2PATE. *Projet de Développement Urbain Intégré et de Résilience (PRODUIR): Cadre de Politique de Réinstallation (CPR)*; M2PATE: Antananarivo, Madagascar, 2018.
46. Rambinintsoa, T. *Les Contraintes Hydrauliques de L'urbanisation d'Antananarivo*. Actes du séminaire International sur le développement urbain, Hôtel de Ville Antananarivo, Madagascar, March 26–30, 2012; Commune Urbaine d'Antananarivo: Antananarivo, Madagascar, 2012.
47. SOMEAH. *Développement Des Grandes Lignes D'un Plan Stratégique D'assainissement à L'échelle de L'agglomération D'Antananarivo*; SOMEAH: Antananarivo, Madagascar, 2010.
48. Antoine, P.; Bocquier, P.; Razafindratsima, N.; Roubaud, F. *Biographies de Trois Générations Dans L'agglomération d'Antananarivo: Premiers Résultats de L'enquête BIOMAD98*; CEPED: Paris, France, 2000.
49. Belrose-Huyghues, V. Un Exemple de Syncrétisme Esthétique Au XIXe Siècle: Le Rova de Tananarive d'Andrianjaka à Radama Ier. *Omaly Sy Anio* **1975**, *1–2*, 173–207.
50. Fournet-Guérin, C. Héritage Reconnu, Patrimoine Menacé: La Maison Traditionnelle à Tananarive. *Autrepart* **2005**, *1*, 51–69. [CrossRef]
51. Blanc-Pamard, C.; Ramiarintsoa, H.R. «ANTANANARIVO, Anc. TANANARIVE». Available online: http://www.universalis.fr/encyclopedie/antananarivo-tananarive/ (accessed on 5 August 2019).
52. CUA. *Histoire de La Ville d'Antananarivo*; CUA: Antananarivo, Madagascar, 2014.
53. Esoavelomandroso-Rajaonah, F. Des Rizières à La Ville: Les Plaines de l'ouest d'Antananarivo Dans La Première Moitié Du XXème Siècle. *Omaly SyAnio Rev. Détudes Hist.* **1989**, 321–337.
54. Wachsberger, J.-M. Les Quartiers Pauvres à Antananarivo. *Autrepart* **2009**, *3*, 117–137. [CrossRef]
55. UNESCO. *La Haute Ville d'Antananarivo*; UNESCO: Paris, France, 2016.
56. Fournet-Guérin, C. *Vivre à Tananarive: Géographie Du Changement Dans La Capitale Malgache*; Karthala Editions: Paris, France, 2007.
57. Fremigacci, J. Les Chemins de Fer de Madagascar (1901–1936). *Afr. Hist.* **2006**, *6*, 161–191.
58. Aubry, C.; Ramamonjisoa, J.; Dabat, M.-H.; Rakotoarisoa, J.; Rakotondraibe, J.; Rabeharisoa, L. Urban Agriculture and Land Use in Cities: An Approach with the Multi-Functionality and Sustainability Concepts in the Case of Antananarivo (Madagascar). *Land Use Policy* **2012**, *29*, 429–439. [CrossRef]
59. Defrise, L.; Burnod, P.; Andriamanga, V. *Terres Agricoles de La Ville d'Antananarivo, Une Disparition Inéluctable?* Université d'Antananarivo: Antananarivo, Madagascar, 2017.
60. ARTELIA. *Programme D'amélioration de L'accès à L'eau Potable et à L'assainissement dans le Canal de la Vallée de L'Est de la CUA*; ARTELIA: Antananarivo, Madagascar, 2016.
61. Mberu, B.; Béguy, D.; Ezeh, A.C. Internal migration, urbanization and slums in sub-Saharan Africa. In *Africa's Population: In Search of a Demographic Dividend*; Springer: Berlin, Germany, 2017; pp. 315–332.
62. Pariente, W. Urbanization in Sub-Saharan Africa and the Challenge of Access to Basic Services. *J. Demogr. Econ.* **2017**, *83*, 31. [CrossRef]
63. Awumbila, M. Drivers of Migration and Urbanization in Africa: Key Trends and Issues. *Int. Migr.* **2017**, *7*, 8.
64. Barrios, S.; Bertinelli, L.; Strobl, E. Climatic Change and Rural-Urban Migration: The Case of Sub-Saharan Africa. *SSRN* **2006**. [CrossRef]
65. Hove, M.; Ngwerume, E.T.; Muchemwa, C. The Urban Crisis in Sub-Saharan Africa: A Threat to Human Security and Sustainable Development. *Stability* **2013**, *2*, 1–14. [CrossRef]
66. Mercandalli, S.; Losch, B.; Rapone, C.; Bourgeois, R.; Khalil, C.A. *Migrations Rurales et Nouvelles Dynamiques de Transformation Structurelle En Afrique Subsaharienne*; FAO: Rome, Italy, 2018.
67. Mlambo, V. An Overview of Rural-Urban Migration in South Africa: Its Causes and Implications. *Arch. Bus. Res.* **2018**, *6*, 63–70. [CrossRef]
68. Pech, L.; Lakes, T. The Impact of Armed Conflict and Forced Migration on Urban Expansion in Goma: Introduction to a Simple Method of Satellite-Imagery Analysis as a Complement to Field Research. *Appl. Geogr.* **2017**, *88*, 161–173. [CrossRef]
69. Henry, S.; Schoumaker, B.; Beauchemin, C. The Impact of Rainfall on the First Out-Migration: A Multi-Level Event-History Analysis in Burkina Faso. *Popul. Environ.* **2004**, *25*, 423–460. [CrossRef]
70. Cobbinah, P.B.; Erdiaw-Kwasie, M.O.; Amoateng, P. Africa's Urbanisation: Implications for Sustainable Development. *Cities* **2015**, *47*, 62–72. [CrossRef]
71. Rakotonarivo, A. Vivre Là-Bas, Exister Ici: Absence et Présence Des Migrants Des Hautes Terres de Madagascar. *Espace Popul. Sociétés Space Popul. Soc.* **2011**, 249–263. [CrossRef]
72. De Brauw, A.; Mueller, V.; Lee, H.L. The Role of Rural–Urban Migration in the Structural Transformation of Sub-Saharan Africa. *World Dev.* **2014**, *63*, 33–42. [CrossRef]
73. Diagne, K. Governance and Natural Disasters: Addressing Flooding in Saint Louis, Senegal. *Environ. Urban.* **2007**, *19*, 552–562. [CrossRef]
74. Saghir, J.; Santoro, J. *Urbanization in Sub-Saharan Africa*; Center for Strategic & International Studies Report; CSIS: Washington, DC, USA, 2018.
75. Satterthwaite, D. The Impact of Urban Development on Risk in Sub-Saharan Africa's Cities with a Focus on Small and Intermediate Urban Centres. *Int. J. Disaster Risk Reduct.* **2017**, *26*, 16–23. [CrossRef]

76. Fox, S. The Political Economy of Slums: Theory and Evidence from Sub-Saharan Africa. *World Dev.* **2014**, *54*, 191–203. [CrossRef]
77. Lemanski, C. Augmented Informality: South Africa's Backyard Dwellings as a by-Product of Formal Housing Policies. *Habitat Int.* **2009**, *33*, 472–484. [CrossRef]
78. Rigon, A.; Walker, J.; Koroma, B. Beyond Formal and Informal: Understanding Urban Informalities from Freetown. *Cities* **2020**, *105*, 102848. [CrossRef]
79. Napier, M. Informal Settlement Integration, the Environment and Sustainable Livelihoods in Sub-Saharan Africa. *Counc. Sci. Ind. Res. S. Afr.* **2007**, 30.
80. Jalayer, F.; De Risi, R.; Kyessi, A.; Mbuya, E.; Yonas, N. Vulnerability of built environment to flooding in African cities. In *Urban Vulnerability and Climate Change in Africa*; Springer: Berlin, Germany, 2015; pp. 77–106.
81. MAHTP; JICA. *Plan D'urbanisme Directeur de L'agglomération d'Antananarivo*; MAHTP: Antananarivo, Madagascar, 2019.
82. Godinot, X.; Razanatsimba, A.; Razafindrasoa, M.; Razanakoto, S.; Ilboudo, M.; Laffitte, A.; Randrianarindiana, P.V.; Tsimihevy, G.; Malakia, J.; Rambelo, D.; et al. *Le Défi Urbain à Madagascar, Quand la Misère Chasse la Pauvreté*; Editions Quart Monde: Montreuil, France, 2012.
83. Ranaivoarimanana, N. Urbanisme de Coalition: Articulation Entre Infrastructures Routières et Plus-Value Foncière Dans La Fabrique Urbaine: Le Cas de La Ville de Tananarive (Madagascar). Ph.D. Thesis, University of Paris-Est, Champs-sur-Marne, France, 2017.
84. Abunyewah, M.; Gajendran, T.; Maund, K. Profiling Informal Settlements for Disaster Risks. *Procedia Eng.* **2018**, *212*, 238–245. [CrossRef]
85. De Risi, R.; Jalayer, F.; De Paola, F.; Iervolino, I.; Giugni, M.; Topa, M.E.; Mbuya, E.; Kyessi, A.; Manfredi, G.; Gasparini, P. Flood Risk Assessment for Informal Settlements. *Nat. Hazards* **2013**, *69*, 1003–1032. [CrossRef]
86. Adelekan, I.; Johnson, C.; Manda, M.; Matyas, D.; Mberu, B.; Parnell, S.; Pelling, M.; Satterthwaite, D.; Vivekananda, J. Disaster Risk and Its Reduction: An Agenda for Urban Africa. *Int. Dev. Plan. Rev.* **2015**, *37*, 33–44. [CrossRef]
87. Alou, A.A.; Lutoff, C.; Mounkaila, H. Relocalisation Préventive Suite à La Crue de Niamey 2012: Vulnérabilités Socio-Économiques Émergentes et Retour En Zone Inondable. *Cybergeo Eur. J. Geogr.* **2019**. [CrossRef]
88. Cambrézy, L. Réfugiés et Migrants En Afrique: Quel Statut Pour Quelle Vulnérabilité? *Rev. Eur. Migr. Int.* **2007**, *23*, 13–28. [CrossRef]
89. iMDC. *Rapport Mondial Sur Le Déplacement Interne*; iMDC: Geneva, Switzerland, 2019; p. 159.
90. Stal, M. Flooding and Relocation: The Zambezi River Valley in Mozambique. *Int. Migr.* **2011**, *49*, e125–e145. [CrossRef]
91. Douglas, I.; Alam, K.; Maghenda, M.; Mcdonnell, Y.; McLean, L.; Campbell, J. Unjust Waters: Climate Change, Flooding and the Urban Poor in Africa. *Environ. Urban.* **2008**, *20*, 187–205. [CrossRef]
92. Bekker, S.; Fourchard, L. *Governing Cities in Africa: Politics and Policies*; HSRC Press: Cape Town, South Africa, 2013.
93. Ravalihasy, T. Habitat et Développement Local: Etude Du Processus de Peuplement et d'occupation de l'espace Dans Les Quartiers d'Andohatapenaka I, II et III. Mémoire de fin D'études, Université d'Antananarivo: Faculté de Droit, d'Économie, de Gestion et de Sociologie, 2005. Thèses Malgaches en Ligne. Available online: http://biblio.univ-antananarivo.mg/theses2/accueil.jsp (accessed on 29 September 2020).
94. Fournet-Guérin, C. Vivre à Tananarive. Crises, Déstabilisations et Recompositions d'une Citadinité Originale. Ph.D. Thesis, Paris-Sorbonne University, Paris, France, 2002.
95. Andriamamonjisoa, S.N.; Hubert-Ferrari, A. Combining Geology, Geomorphology and Geotechnical Data for a Safer Urban Extension: Application to the Antananarivo Capital City (Madagascar). *J. Afr. Earth Sci.* **2019**, *151*, 417–437. [CrossRef]
96. Förster, T.; Ammann, C. Les Villes Africaines et Le Casse-Tête Du Développement. Acteurs et Capacité d'agir Dans La Zone Grise Urbaine. *Int. Dev. Policy Rev. Int. Polit. Dév.* **2018**, *10*. [CrossRef]
97. Fox, S. Urbanization as a Global Historical Process: Theory and Evidence from Sub-Saharan Africa. *Popul. Dev. Rev.* **2012**, *38*, 285–310. [CrossRef]
98. Fox, S. *Understanding the Origins and Pace of Africa's Urban Transition*; Crisis States Research Centre Working Paper (Series 2) No. 98 (89); London School of Economics and Political Science: London, UK, 2011.
99. Banque Mondiale Population Urbaine. Available online: https://donnees.banquemondiale.org/indicator/SP.URB.TOTL.IN.ZS (accessed on 22 August 2020).
100. Amoako, C.; Inkoom, D.K.B. The Production of Flood Vulnerability in Accra, Ghana: Re-Thinking Flooding and Informal Urbanisation. *Urban Stud.* **2018**, *55*, 2903–2922. [CrossRef]
101. Cohen, B. Urban Growth in Developing Countries: A Review of Current Trends and a Caution Regarding Existing Forecasts. *World Dev.* **2004**, *32*, 23–51. [CrossRef]
102. Aldeghiri, M. Les Cyclones de Mars 1959 à Madagascar. In *Annuaire Hydrologique de la France d'Outre-Mer: Année 1957*; ORSTOM: Paris, France, 1959.
103. MDAT; UN-Habitat; UNDP. *Politique Nationale de l'Aménagement Du Territoire (PNAT)*; MDAT: Antananarivo, Madagascar, 2006.
104. Abass, K.; Adanu, S.K.; Agyemang, S. Peri-Urbanisation and Loss of Arable Land in Kumasi Metropolis in Three Decades: Evidence from Remote Sensing Image Analysis. *Land Use Policy* **2018**, *72*, 470–479. [CrossRef]
105. Olisoa, F.R. Mutations Des Espaces Périurbains d'Antananarivo: Population, Habitat et Occupation Du Sol. Ph.D. Thesis, Strasbourg University, Strasbourg, France, 2012.

106. Rabemalanto, N. Vulnérabilité Résidentielle Des Ménages et Trappes à Pauvreté En Milieu Urbain. Les" Bas-Quartiers" d'Antananarivo. Ph.D. Thesis, University of Paris-Saclay, Gif-sur-Yvette, France, 2018.
107. Morisset, J.; Andriamihaja, N.A.; Randriamiarana, Z.B.; Graftieaux, P.; Rakotoniaina, P.; Andriamihamina, H.D.; Daussin, A.; Struyven, D.; Faure, C. *L'urbanisation Ou Le Nouveau Défi Malgache*; Banque Mondiale: Washington, DC, USA, 2011.
108. MAEP. *Monographie de La Région d'Antananarivo*; MAEP: Antananarivo, Madagascar, 2003.

Article

Flood Suspended Sediment Transport: Combined Modelling from Dilute to Hyper-Concentrated Flow

Jaan H. Pu [1,*], Joseph T. Wallwork [1], Md. Amir Khan [2], Manish Pandey [3], Hanif Pourshahbaz [4], Alfrendo Satyanaga [5], Prashanth R. Hanmaiahgari [6] and Tim Gough [1]

1   Faculty of Engineering and Informatics, University of Bradford, Bradford DB7 1DP, UK; theowallwork@gmail.com (J.T.W.); t.gough@bradford.ac.uk (T.G.)
2   Galgotias College of Engineering and Technology, Greater Noida, Uttar Pradesh 201310, India; amirmdamu@gmail.com
3   National Institute of Technology (NIT), Warangal, Telangana 506004, India; mpandey@nitw.ac.in
4   Department of Civil Engineering, University of Zanjan, Zanjan 45371-38791, Iran; H.pourshahbaz@znu.ac.ir
5   Department of Civil and Environmental Engineering, Nazarbayev University, Nur-Sultan 010000, Kazakhstan; alfrendo.satyanaga@nu.edu.kz
6   Department of Civil Engineering, Indian Institute of Technology, Kharagpur 721302, India; hpr@civil.iitkgp.ac.in
*   Correspondence: j.h.pu1@bradford.ac.uk

Citation: Pu, J.H.; Wallwork, J.T.; Khan, M..A.; Pandey, M.; Pourshahbaz, H.; Satyanaga, A.; Hanmaiahgari, P.R.; Gough, T. Flood Suspended Sediment Transport: Combined Modelling from Dilute to Hyper-Concentrated Flow. *Water* **2021**, *13*, 379. https://doi.org/10.3390/w13030379

Academic Editor: Jorge Leandro and James Shucksmith
Received: 13 December 2020
Accepted: 28 January 2021
Published: 1 February 2021

Publisher's Note: MDPI stays neutral with regard to jurisdictional claims in published maps and institutional affiliations.

Copyright: © 2021 by the authors. Licensee MDPI, Basel, Switzerland. This article is an open access article distributed under the terms and conditions of the Creative Commons Attribution (CC BY) license (https://creativecommons.org/licenses/by/4.0/).

Abstract: During flooding, the suspended sediment transport usually experiences a wide-range of dilute to hyper-concentrated suspended sediment transport depending on the local flow and ground conditions. This paper assesses the distribution of sediment for a variety of hyper-concentrated and dilute flows. Due to the differences between hyper-concentrated and dilute flows, a linear-power coupled model is proposed to integrate these considerations. A parameterised method combining the sediment size, Rouse number, mean concentration, and flow depth parameters has been used for modelling the sediment profile. The accuracy of the proposed model has been verified against the reported laboratory measurements and comparison with other published analytical methods. The proposed method has been shown to effectively compute the concentration profile for a wide range of suspended sediment conditions from hyper-concentrated to dilute flows. Detailed comparisons reveal that the proposed model calculates the dilute profile with good correspondence to the measured data and other modelling results from literature. For the hyper-concentrated profile, a clear division of lower (bed-load) to upper layer (suspended-load) transport can be observed in the measured data. Using the proposed model, the transitional point from this lower to upper layer transport can be calculated precisely.

Keywords: parameterised power-linear model; hyper concentration; dilute concentration; suspended sediment transport; flood; sediment size parameter; rouse number; mean concentration; flow depth

1. Introduction

Sediment transport is a common phenomenon during flooding. When sufficient lift force on sediment particles exists to overcome the frictional grips in between them, flow turbulence especially in the upward direction will generate sediment suspension [1,2]. Unlike the bed load, this suspended load is still not well-understood especially for those sediments with highly soluble behaviour in flow [3].

Two-phase flow is usually subjected to complex mixture between the solid and fluid phases. It is complex to mathematically model, in particular when one considers the natural flow in compound or irregular channels such as those studied by Pu [4] and Pu et al. [5]. Some models [6–8] resolve these complexities by neglecting turbulence and forces acting on the sediment particle surfaces, such as the effects of turbulent diffusion in laminar uniform flow or particle–particle collisions within dilute flows. However, applying these assumptions significantly hinders the modelling accuracy. As a result, recent studies have attempted to incorporate the resultant lift and drag forces acting on the particle phase [9–11], and this has resulted in diverse formulations for predicting the suspended sediment profile within the flow.

The general consensus when modelling the sediment-laden flow is to assume a representative two-dimensional plane due to the complexity of full 3D modelling [12]. The sediment concentration is normally considered to change with height from the bed [13]; and various flow parameters can be incorporated into mathematical models to determine the full concentration profile. These parameters commonly include: particle fall velocity, particle diameter, Rouse number and mean concentration [14,15]. Within the field of sediment profiling a range of mathematical concepts has been adopted to predict the concentration profile. Goree et al. [16] used continuum theory and incorporated the effect of drift flux due to flow turbulence implemented using large-eddy simulation. However, it was found that the computed results were less accurate in the near-wall region (also agree with [17,18]). Rouse [6] proposed diffusion theory to form one of the simplest mathematical approaches. Despite the apparent simplicity, this diffusion theory-based calculation gave reasonable and efficient prediction of the suspended solid behaviour and has subsequently been utilised as the basis for many further studies.

Another commonly used mathematical concept is that of kinetic theory. This theory is widely regarded as one of the most precise approaches to model sediment concentration distribution as it includes the response of both the solid and liquid phases as well as the interactions between them [11]. Other theories have also been produced and shown to give reasonable results, such as the combination between kinetic and diffusion theories proposed by Ni et al. [19].

In this paper, we are motivated to seek a representative model to analytically calculate the suspended sediment transport profile, since currently there is a lack of such modelling in literature to inclusively represent the diluted, transitional, and dense suspended sediment transport. In the view of this research gap, in this study, the reported models are analysed and prominent flow parameters are assessed. A method of parameterisation is introduced using an analytical regression analysis technique. Consequently, separate parameterised expressions have been proposed for a wide range of flow conditions (i.e., from dilute to hyper-concentrated flow), before being adopted into a coupled power-linear concentration model. Various tests have also been conducted to validate the proposed model with published experimental data to assess the model's accuracy.

2. Models Review

Diffusion theory has played an important role in mathematical modelling of the suspended solid transport and has been used as the basis of the models by van Rijn [20], Wang and Ni [15], McLean [21], and Zhong et al. [22]. Rouse [6] derived his model from Fick's Law, which defines diffusion theory, and states that diffusion from an area of high concentration to an area of low concentration should be balanced by the product of the settling velocity and concentration as described in Equation (1) [23]:

$$D\frac{dc}{dy} = -\omega_0 c \tag{1}$$

where D is sediment diffusivity (m^2/s), c is concentration (dimensionless), ω_0 is settling velocity (m/s), and y represents the vertical space across a flow depth (m). Within Fick's formula, assumptions about sediment diffusivity must be made, for which Rouse proposed that the upward diffusion was a result of the vertical flux due to turbulence and assumed that the suspended particles was only associated with fluid turbulence diffusivity [7]. This agrees with the law of wall such that the sediment diffusivity is defined by the shear velocity u_*. Therefore, D can be defined by (Equation (2)):

$$D = \kappa y u_* (1 - \varepsilon) \tag{2}$$

where κ is von Karman constant (dimensionless), u_* is the shear velocity (m/s) and ε is the characteristic height (dimensionless) defined as the vertical distance, y, from the boundary normalised by the flow depth h (Equation (3)):

$$\varepsilon = \frac{y}{h} \qquad (3)$$

Hence ε is limited by $0 < \varepsilon \leq 1$.
Inserting Equation (2) into Equation (1) gives Equation (4) as follow:

$$\frac{1}{c}dc = -\frac{\omega_0}{\kappa y u_*(1-\varepsilon)}dy \qquad (4)$$

Integrating Equation (4) between the boundaries ε and a reference characteristic height ε_a gives the Rouse formula (Equation (5))

$$\frac{c}{c_a} = \left[\frac{1-\varepsilon}{\varepsilon} \cdot \frac{\varepsilon_a}{1-\varepsilon_a}\right]^{\frac{\omega_0}{\kappa u_*}} \qquad (5)$$

where c_a is the concentration at the reference height (dimensionless). ε_a is described as the point where suspended load transport begins to take place and suggested to be 0.005 by Hsu et al. [7]. Under the assumption made by Rouse, the concentration distribution profile becomes more uniform with decreasing Rouse number which can be achieved by using sediment with low settling velocity or by increasing shear velocity, where the Rouse number P can be described by Equation (6):

$$P = \frac{\omega_0}{\kappa u_*} \qquad (6)$$

A modified model from Rouse has been presented by Kundu and Ghoshal [14] in which they recognised that the sediment concentration distribution can follow more than one profiles, as depicted in Figure 1. The most common profile (Type I) shows a monotonic decrease in concentration with height, and it happens when the flow concentration is dilute. The Type II profile shows an increase in concentration with height to a peak value above the bed, thereafter the concentration decreasing with height (it happens when flow is experiencing transitional concentration between dilute and dense condition). This Type II profile gives rise to a transitional point splitting the distribution into an upper flow region (above maximum concentration) and a lower flow region (below the maximum concentration in the near-bed region). The Type III profile occurs when the flow is subjected to hyper-concentration of sediment and exhibits a steady increase from the bed followed by a decrease in concentration towards the outer region of the flow.

In terms of modelling, Type I allows the most simplistic solution as it can be fitted using the common Rouse approach. However, the heavy sediment-laden flows usually present Type II or III profile. In common with the Rouse model, the dependent variable for the model presented by Kundu and Ghoshal [14] is ε, where its functions can be defined as (Equations (7) and (8))

$$\varphi_1 = b_1 \varepsilon^{\alpha_1} + q_1 \qquad (7)$$

and,

$$\varphi_2 = b_2 \varepsilon^{\alpha_2} + q_2 \qquad (8)$$

in which $b_1, \alpha_1, q_1, b_2, \alpha_2$ and q_2 are empirical coefficients to be determined from experimental data.

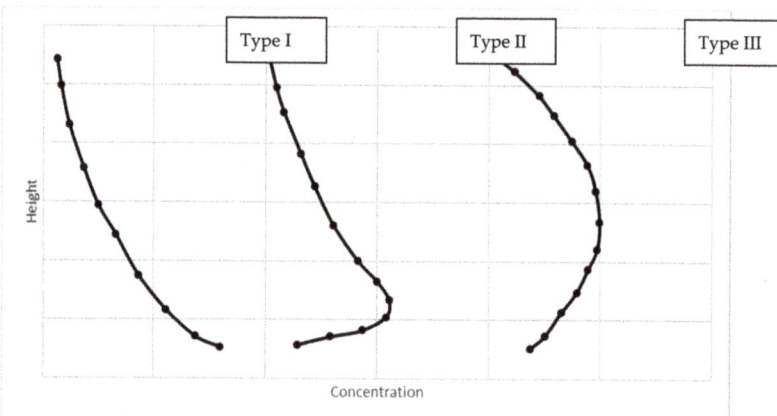

Figure 1. Type I, II and III Concentration Profiles.

Using the asymptotic matching technique by Almedeij [24], the concentration profile for both adjacent sections can be expressed in the Equation (9):

$$\frac{c}{\bar{c}} = \frac{1}{(b_1 \varepsilon^{\alpha_1} + q_1)^{-1} + (b_2 \varepsilon^{\alpha_2} + q_2)^{-1}} \quad (9)$$

within which φ_1 represents the lower suspension flow region and φ_2 represents the upper suspension region. Using this technique, Kundu and Ghoshal [14] produced the empirical coefficients by calibration with previously published experimental data.

Several experimental studies (i.e., Einstein and Qian [25], Bouvard and Petkovic [26], Wang and Ni [15]) have also shown that the sediment profile follows a power law solution within the dilute-concentrated flow regime. This can be described by Equations (7) and (8) through the following simplification in Equation (10):

$$\varphi = b\varepsilon^{\alpha} \quad (10)$$

where it is formed when the parameter q in Equations (7) and (8) is set as zero to produce a power law solution. In this formulation, Equation (10) reverts to a similar form as the Rouse formula shown in Equation (5).

However, within extreme flow conditions such as hyper-concentrated flow where the sediment profile has been proven to deviate from the power law distribution. Experimental results yield a linear profile due to an increase in particle–particle interactions. Equation (7) to (8) should thus take a form of the following in Equation (11):

$$\varphi = b\varepsilon + q \quad (11)$$

with exponent α equals to unity.

Limitations of this Rouse-type formulation have been evidenced by the measurements of Sumer et al. [27], Greimann and Holly [9], Jha and Bombardelli [10], Kironoto and Yulistiyanto [28], and Goeree et al. [16]. Owing to its derivation from diffusion theory, the Rouse formula provides a single-phase approach focusing on the sediment particles. As a result, the Rouse formula is limited to the representations of flows exhibiting Type I concentration profiles (Figure 1). Due to the boundary assumptions of the Rouse formula, the resultant concentration profile must always revert to zero at the fluid surface and infinity at the bed [29]. Huang et al. [8] further stated that the Rouse formula can lose its accuracy near-bed particularly when dealing with high boundary roughness. One of the attempts to improve the Rouse model is to incorporate an additional factor β into the

Rouse number producing a damping effect, where β is the coefficient of proportionality for the diffusion coefficient for sediment transfer [29].

Greimann and Holly [9] derived a formula using a two-phase approach to the Rouse model. Within their study it is highlighted that, due to Rouse's lack of consideration of particle–particle interactions, the Rouse formula is only valid when $c < 0.1$. As the Rouse formula is derived from Fick's law, it is only applicable to flow when the bulk Stokes number S_b (which is a parameter commonly used to define characteristic of suspended particles in a fluid flow) is very small such that the fluid and solid phases are transported almost in equilibrium. Therefore, it can be concluded that while the Rouse formula gives reasonable calculation to sediment profiling, it is limited by the absence of mechanical forces such as particle–particle interactions and particle inertia, and by its lack of effective sediment parameterisation, i.e., related to sediment size. In comparison, the models proposed by Wang and Ni [15], Ni et al. [19], and Zhong et al. [22] utilised either exponential or power laws to precisely represent suspended sediment profiles across the whole flow depth and with a variety of concentration levels. They adapted kinetic concepts for considering the particle concentration, thus can model two-phase interactions. Additionally, they used empirical fit to determine the profile characteristic, and identified various flow and sediment parameters that can be potentially used to define the concentration profile.

The aim of this study is to investigate the relationship between various flow and sediment parameters to form an improved representation to Equations (7)–(9). This will form a parameterised expression of final suspended particle characteristic model and allow an effective prediction of its concentration profile. The flow parameters to be investigated are Rouse number P, size parameter S_z, and mean concentration $\bar{c}$. Additionally, this kind of formulation using the parameterised expressions to improve the suspended sediment transport modelling has so far not been explored in other studies, hence this investigation is crucially needed to study the performance of such modelling.

3. Proposed Modelling

Many studies have investigated the relevant parameters for considering a concentration profile, including the Rouse number (defined in Equation (6)), particle size, mean concentration, and flow depth [6,27,30,31]. By referring to Equation (9), the variables are related to the coefficients of power-linear law as follows in Equation (12):

$$b_1, b_2, \alpha_1, \alpha_2, q_1, q_2 = f(P, S_Z, \bar{c}) \tag{12}$$

where, S_z is the dimensionless size parameter ($S_z = d/h$), in which d is the sediment particle diameter and h is the flow depth. In this investigation, we collected data from various reported experimental studies (as detailed in Table 1) to inspect the distribution of each power-linear law coefficient toward the physical parameters of Rouse number, particle size, and mean concentration, and to deduce a modified Rouse model for validation tests. It can be observed from Table 1 that the utilised data sources are in a wide range. In particular, the $\bar{c}$ range in the utilised literature are ranging from 0.00013 to 0.147, which giving a thorough test of concentrations from dilute to hyper-concentrated flow conditions.

Table 1. Data sources for parameterised modelling.

Data Sources	h (cm)	d (mm)	ω_0 (cm/s)	u_* (cm/s)	$\bar{c}(\times 10^{-3})$
Bouvard and Petkovic [26]	7.5	2.00–9.00	1.81–2.70	2.54–5.41	2.1–4.5
Cellino and Graf [32]	12.0	0.135	1.20	4.30–4.50	96–147
Coleman [30]	17.0–17.4	0.21–0.42	1.23–1.31	4.10	0.13–0.28
Muste et al. [33]	2.1	0.21–0.25	0.06	4.00–4.30	0.46–1.62

3.1. Rouse Number

Two of the main parameters affecting drag on a sediment particle are the settling and shear velocities. A dimensionless form of these parameters together with von Karman constant is the Rouse number as defined in Equation (6). By studying each parameter in Equation (9) against P, we can produce Figures 2–5 below.

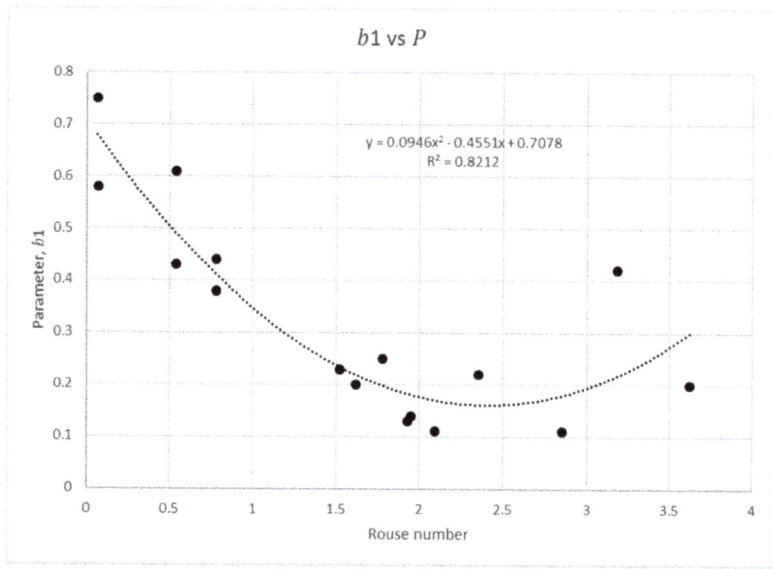

Figure 2. Rouse number regression analysis for coefficient b_1.

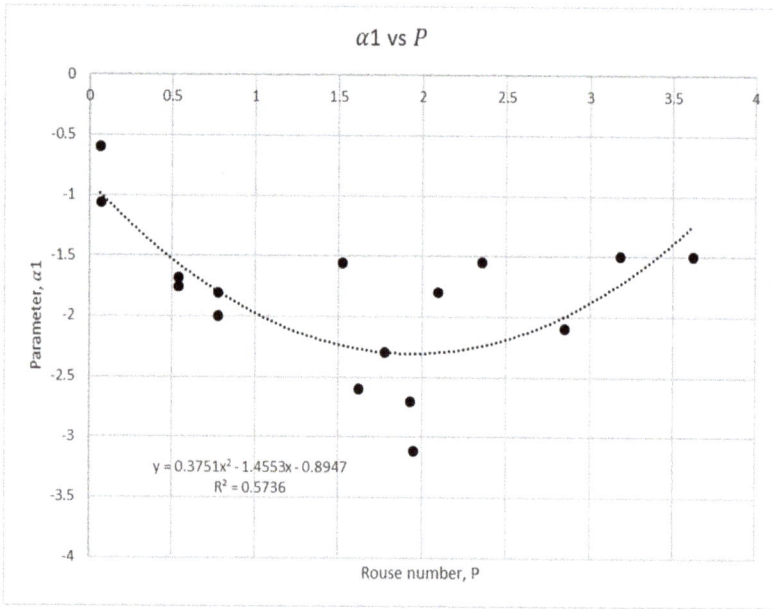

Figure 3. Rouse number regression analysis for coefficient α_1.

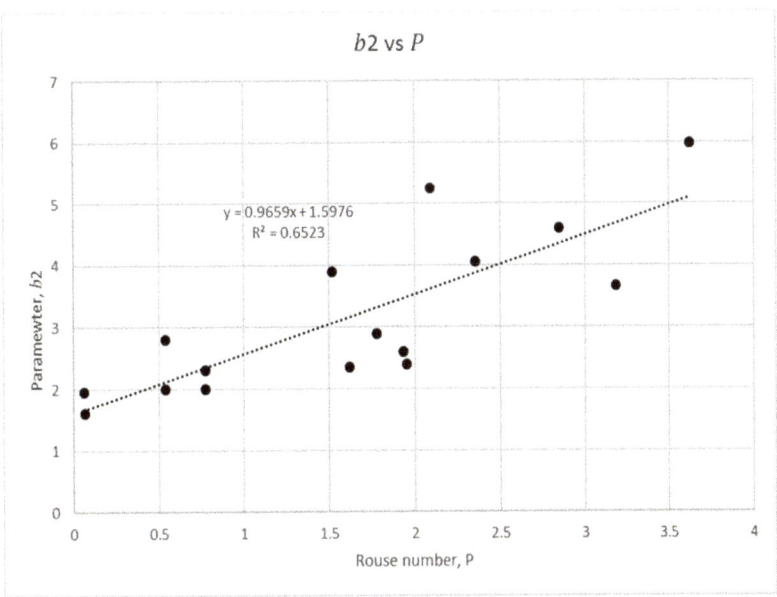

Figure 4. Rouse number regression analysis for coefficient b_2.

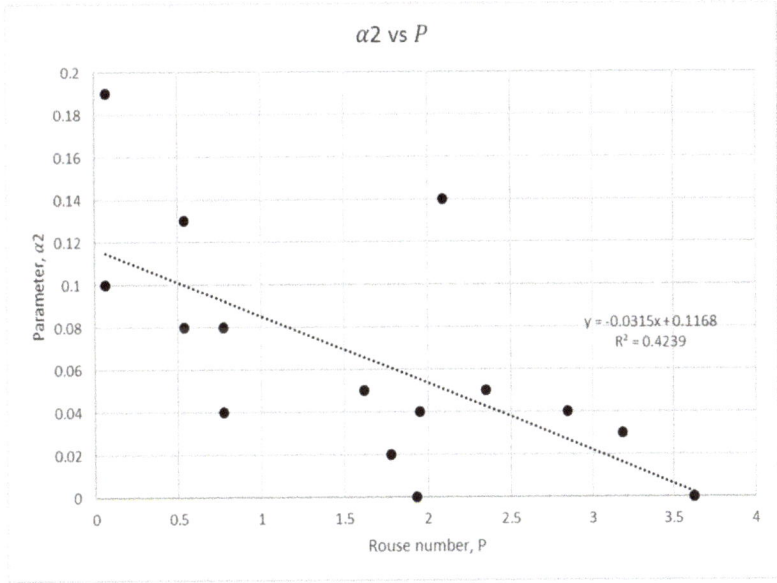

Figure 5. Rouse number regression analysis for coefficient α_2.

Figures 2–5 show that there is a quadratic relationship for the parameters b_1 and α_1, and a linear relationship for b_2 and α_2 against P. The regression analysis shows all coefficients have $R^2 > 0.5$, except for α_2. This finding shows P provides reasonable fit to be represented by power-law and its coefficients in wide range of measured data. As analysed by Kundu and Ghoshal [11], the Rouse number function does not provide a good representation to the hyper-concentrated profile, and hence the hyper-concentrated flow data have been omitted in Figures 2–5.

3.2. Size Parameter

Particle size is another factor that significantly affects sediment drag and lift. The surface area of a particle, determined by the diameter for a spherical particle, can affect the effectiveness of interactive contacts act on the particle. Additionally, particle diameter also influences its settling velocity [34]. In Figures 6–9, the dimensionless S_Z is plotted against the proposed model's power-law coefficients. In the original Rouse approach [6] or modified Rouse model (as used in Kundu and Ghoshal [14]), the effect of particle size has not been considered, even though it is a crucial factor in determining the suspended sediment behaviour.

In Figures 6–9, it can be observed that a quadratic relationship describes the variation of b_1 and b_2 with S_Z while a logarithmic relationship for α_1 and α_2 against S_Z is observed. All the figures show R^2 regression lower than 0.5 with the exception of b_1. This low regression shows that α and b are harder to be represented by S_Z, which in turns exposes the difficulty of modelling using S_Z. Its analysis further suggests that the particle size factor is harder to be fixed. In the analytical modelling studies of Wang and Ni [15] and Ni et al. [19], the particle diameter has been fitted by using a coefficient in the concentration equation extracted from the measured data. The described tests have shown that it is hard to capture the characteristic of concentration profiles when different sediment diameters have been tested. This further affirms the difficulty of finding a representative function for the particle size parameter investigated here.

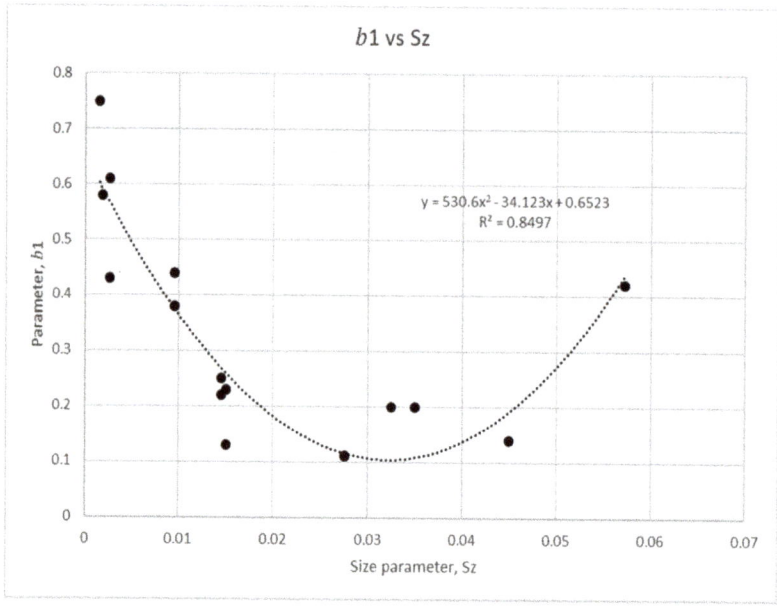

Figure 6. Size parameter regression analysis for coefficient b_1.

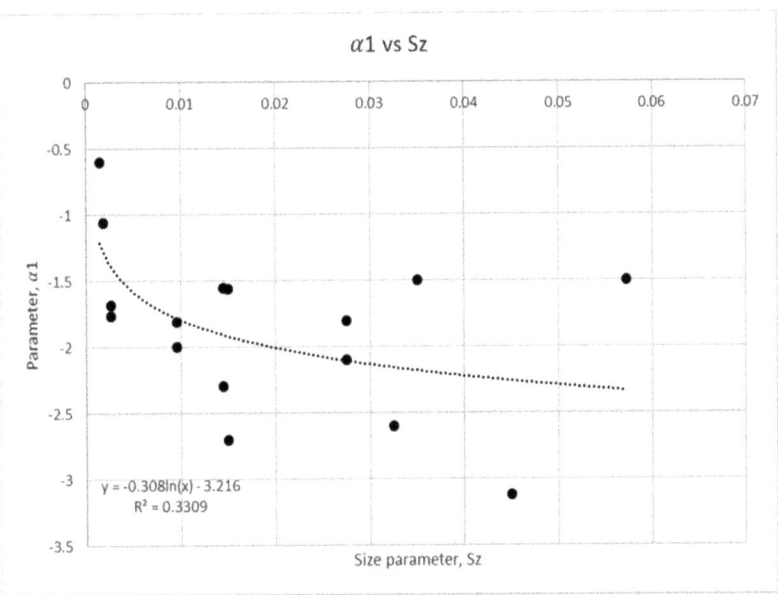

Figure 7. Size parameter regression analysis for coefficient b_2.

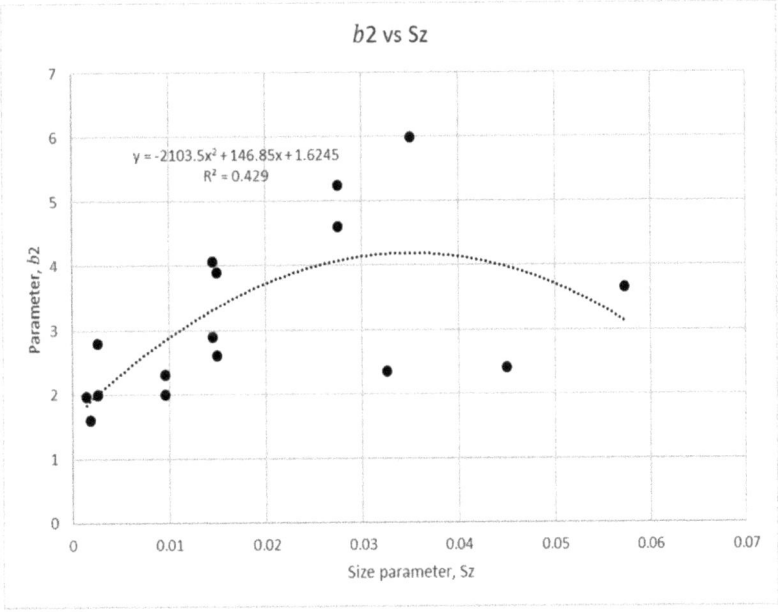

Figure 8. Size parameter regression analysis for coefficient b_2.

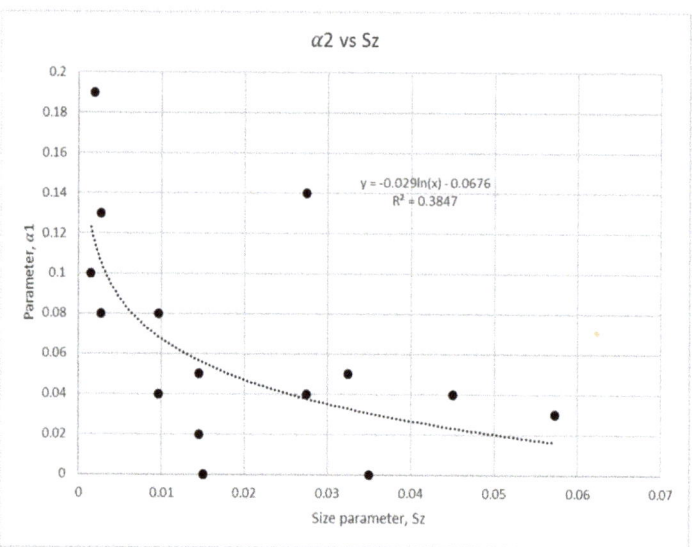

Figure 9. Size parameter regression analysis for coefficient α_2.

3.3. Mean Concentration

Michalik [35] and Cellino and Graf [32] investigated the measured sediment concentration profiles for hyper-concentrated flows. Their results showed that for such flows, the sediment profile follows a more linear distribution as opposed to the common power law observed in other dilute flow studies. Another empirical observation by Machalik [35] was that the mean concentration has key dominant impact on the characteristic of concentration distribution compared to Rouse number or particle size. Hence this study will formulate the analytical approach by mean concentration for use in the linear law modelling. Regression analysis was also used to identify the fit between the linear law coefficients and mean concentration, as presented in Figures 10–13.

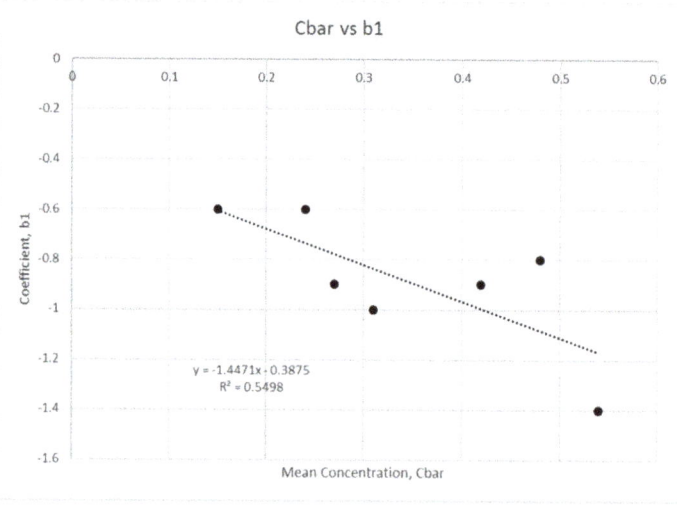

Figure 10. Mean concentration regression analysis for coefficient b_1.

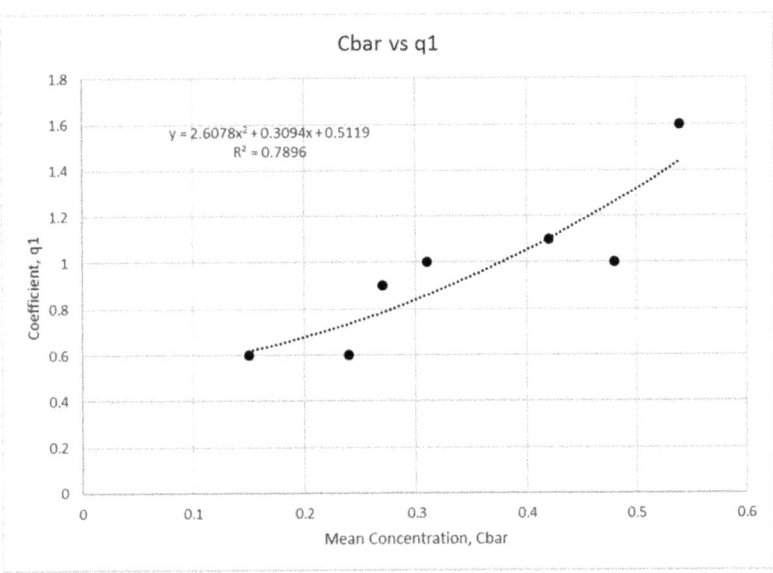

Figure 11. Mean concentration regression analysis for coefficient q_1.

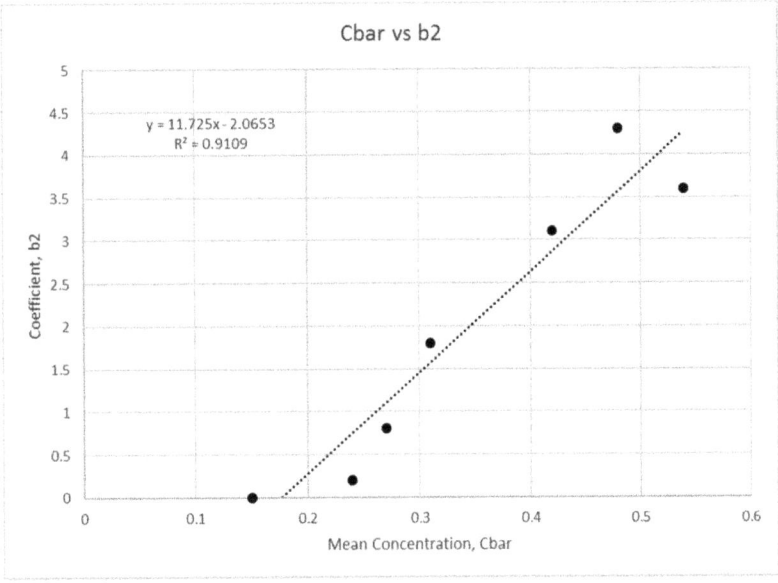

Figure 12. Mean concentration regression analysis for coefficient b_2.

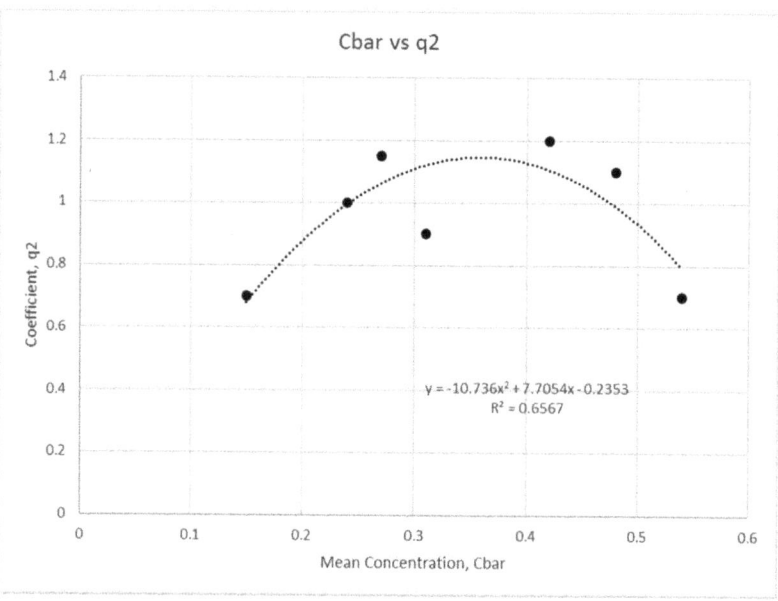

Figure 13. Mean concentration regression analysis for coefficient q_2.

The results show that a linear fit describes the variation for the coefficients b_1 and b_2 with mean concentration; while a quadratic relationship describes the variation of the coefficients q_1 and q_2 with $\bar{c}$. The fits show an R^2 regression higher than 0.5 without exception. This finding evidences a clear correlation between $\bar{c}$ and the hyper-concentrated profile.

3.4. Hyper-To-Dilute Boundary

The findings from the above sections are adapted into the model of Equation (9) to form a parameterised expression for the sediment concentration distribution across the flow depth. The proposed sediment concentration calculative model is governed by a coupled approach. A power law is utilised to represent the dilute sediment concentration (when $0 < \bar{c} < 0.1$), whereas a linear law is used for the dense hyper-concentration (when $\bar{c} \geq 0.1$). This hyper-to-dilute boundary has been set by benchmarking the investigation of Greimann and Holly [9] on dilute flow definition and Rouse model limit. Our proposed coefficients found from the above sections can be represented as:

For the dilute regime, where $0 < \bar{c} < 0.1$ (Equations (13)–(17)):

$$q_1 = q_2 = 0 \tag{13}$$

$$b_1 = 0.047P^2 - 0.23P + 270S_z^2 - 17S_z + 0.68, \tag{14}$$

$$\alpha_1 = 0.19P^2 - 0.073P - 0.15\ln(S_z) - 2.0, \tag{15}$$

$$b_2 = 0.48P - 1100S_z^2 + 73S_z + 1.6, \tag{16}$$

$$\alpha_2 = -0.016P - 0.015\ln(S_z) + 0.025. \tag{17}$$

For the hyper-concentrated regime where $\bar{c} \geq 0.1$ (Equations (18)–(22)):

$$\alpha_1 = \alpha_2 = 1.0 \tag{18}$$

$$b_1 = -1.4\,\bar{c} - 0.39, \tag{19}$$

$$q_1 = 2.6\bar{c}^2 + 0.39\bar{c} + 0.51, \tag{20}$$

$$b_2 = 11\,\bar{c} - 2.1, \tag{21}$$

$$q_2 = -11\,\bar{c}^2 + 7.7\,\bar{c} - 0.24. \tag{22}$$

4. Model Validations

The model presented within this paper is validated against the experimental data of Wang and Ni [31], Wang and Qian [36], and Michalik [35]. It has also been compared with the previously proposed models by Wang and Ni [31], Ni et al. [19], and Zhong et al. [22]. Wang and Ni [31] presented a theoretical distribution model derived from the kinetic theory. Their model is limited to dilute flow and therefore predicts Type I and limited Type II profiles only. In their assumption, the particle interaction has been neglected, and as a result, they attributed the classification of distribution profile solely to the fluid-induced lift forces. It is also noteworthy within their study that when particle size is small the distribution tends to follow the Type I profile.

The model proposed by Ni et al. [19] used a fusion of kinetic and continuum theories, where kinetic theory using the Boltzmann equation being applied to the solid-phase and continuum theory to the fluid-phase. Within their derivation, the empirically weighted forces have been used to act upon sediment to represent two-phase interactions. The model has been proposed to be applicable to both dilute and dense flows. The model proposed by Zhong et al. [22] is more complex when compared to the other two above-mentioned models. It is based on a tertiary approach where the model can be simplified under various empirically-driven assumptions. Within their research, the experimental testing covers Type I, II, and III profiles; though due to its complexity, their Type III profile required a dynamic value of the empirical damping function to fit for different flow conditions.

4.1. Wang and Ni

The measured data of Wang and Ni [31] assessed dilute flow within pipes. The concentrations tested were extremely dilute ranging from $0.00042 \leq \bar{c} \leq 0.0033$, where these tested conditions are presented in Table 2. This validation exercise will provide a good test to the proposed model capability to capture extremely dilute flow. The sediments tested were grains and coarse sands with particle diameter ranging from 0.58 mm $\leq d \leq 2.29$ mm. The results are presented in Figure 14A–J. The proposed model shows a reasonable correspondence to the experimental data. The measurements of Wang and Ni [31] show that for dilute flow the sediment concentration tends to follow the Type I or II concentration profile with the maximum concentration occurring in the near-bed region.

Table 2. Data by Wang and Ni [31].

Test No.	d (mm)	ω_0 (cm/s)	u_* (cm/s)	$\bar{c}\,(\times 10^{-3})$
A1	1.80	2.56	3.28	3.30
A3	1.40	6.90	4.76	3.10
A4	1.10	5.15	4.52	2.40
A5	0.58	4.51	4.79	0.97
A6	0.60	3.79	4.90	0.57
A7	2.29	6.15	4.83	0.44
B3	1.40	6.90	6.11	1.98
B4	1.10	5.15	6.15	2.00
B5	0.58	4.51	6.33	1.94
B6	0.60	3.79	6.23	0.42

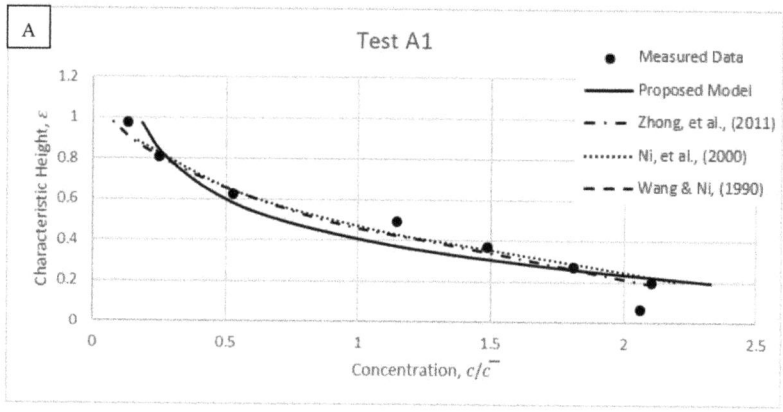
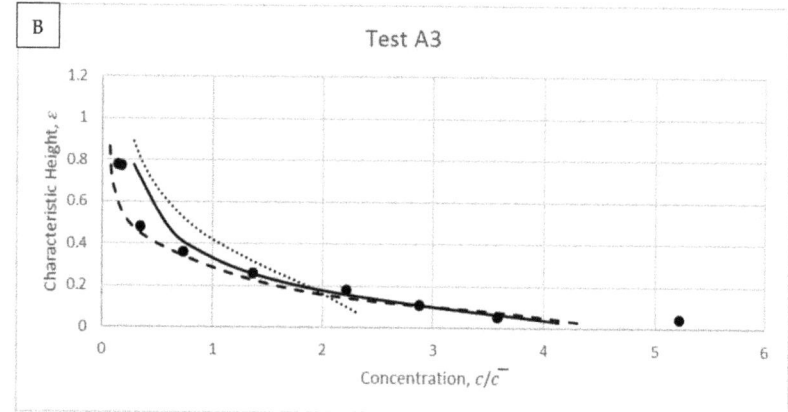
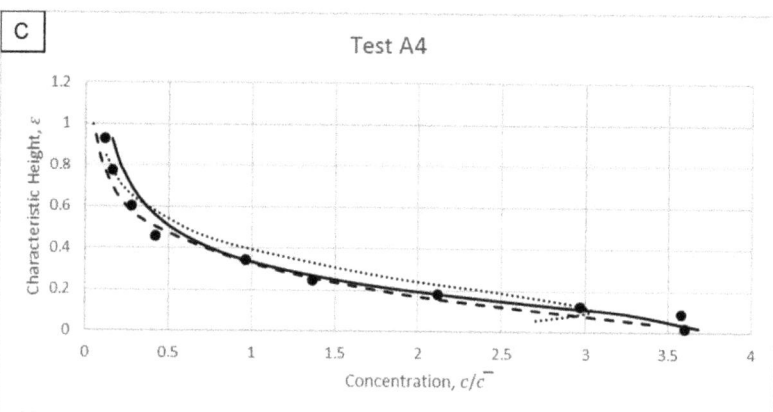

Figure 14. *Cont.*

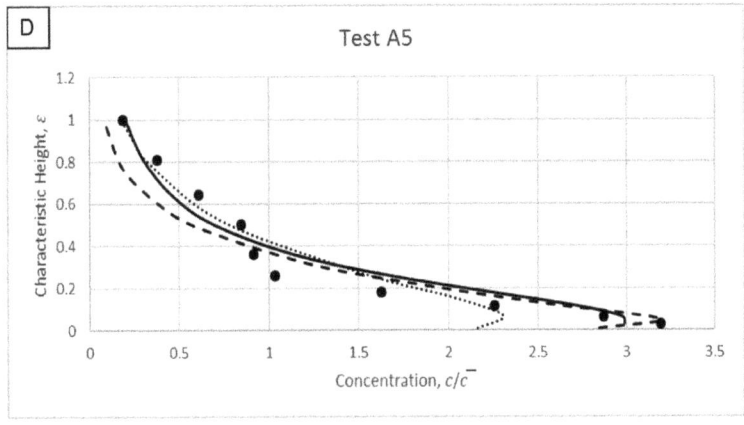
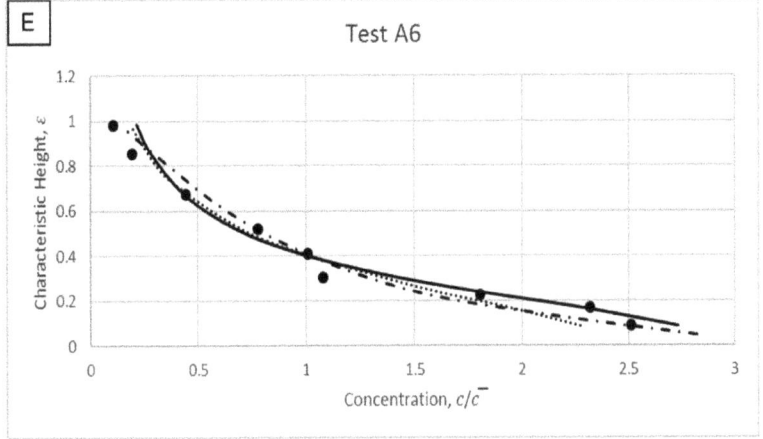
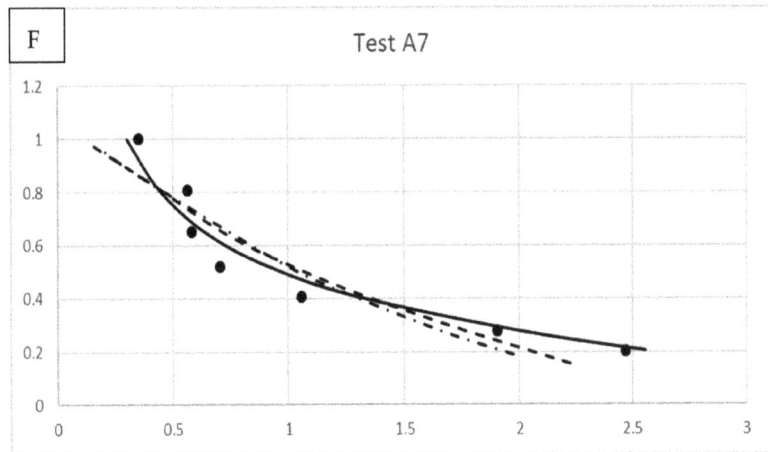

Figure 14. *Cont.*

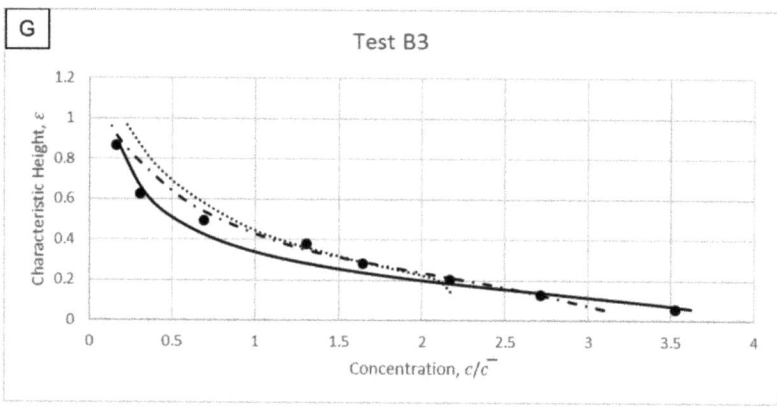
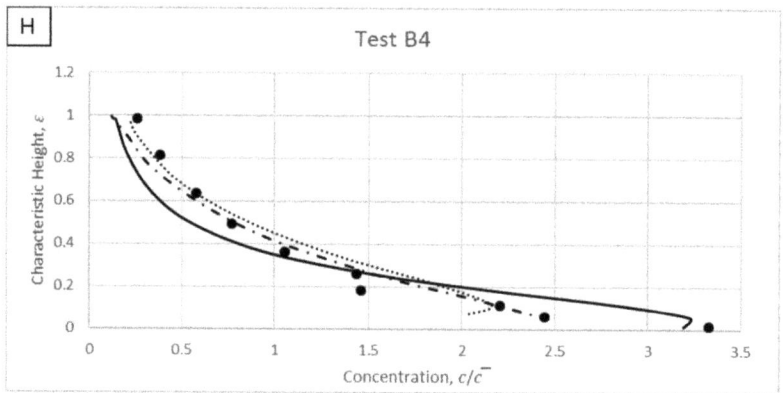
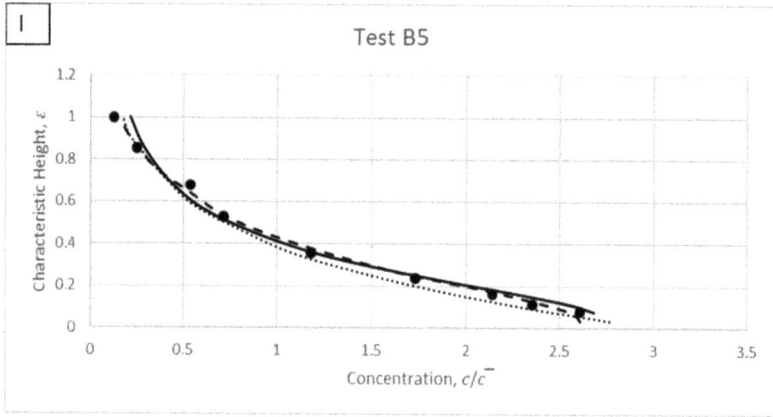

Figure 14. *Cont.*

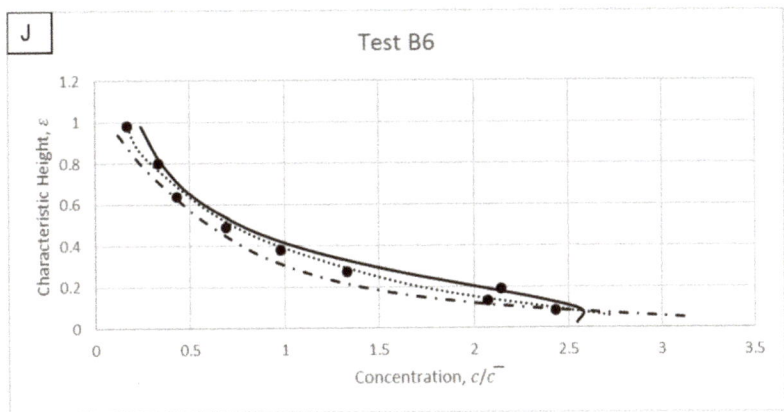

Figure 14. Modelled results and comparisons against experimental data of Wang and Ni [31].

Observation of Figure 14A–J shows that overall there is a better fit by the proposed and other models away from the near-bed region. This is coherent with the suggestions from literature (i.e., Kundu and Ghoshal [11]; Greimann and Holly [9]) stating that the possibility of particle–particle interactions increases at near-bed to produce more challenging conditions for the mathematical modelling.

The experimental data in Figure 14F utilised the largest particles among all the measured data of Wang and Ni [31]. Therefore, larger interaction forces can be expected between the solid–fluid phases due to the larger surface area of each sediment particle. One can observe that the proposed model shows a concentration distribution for Figure 14F which is consistent with the measurements. Compared with the model of Zhong et al. [22], which does not take into account the particle size, the proposed model shows promising computation of big particle measured data.

4.2. Wang and Qian

Wang and Qian [36] studied the effect of dilute to dense concentrations in open channel flow using an experimental recirculating-tilting flume. Their experiments tested a wide range of sediment diameters 0.15 mm $\leq d \leq$ 0.96 mm and concentrations 0.0102 $\leq \bar{c} \leq$ 0.0906 (as shown in Table 3). Their tests are compared against the proposed and other models, with the results presented in Figure 15A–D. Overall, the models show lower accuracy to reproduced measured data throughout the flow depth with increasing mean concentration. The proposed model shows a reasonable fit to the experimental data in the upper flow region. Within this region, the main forces acting on the sediment particles are the lift and drag due to the fluid induced forces and particle inertia. This supports the hypothesis that the proposed model can compute the solid–fluid interactions for particles with reasonable accuracy owing to its inclusion of Rouse and size parameters.

An overview of all the results displayed in Figure 15A–D demonstrates that the accuracy of different models reduces at the lower suspension region, including the proposed model. Within Type II profiles, the local maximum in concentration is observed closer to the bed compared to Type III profiles. Near to the wall boundary, the sediment distribution is governed by the bed-load behaviour as opposed to the suspended load. The plastic particles were used within the flows tested here, and they generate less significant movement than the normal and natural sediment under the acting of particle–particle interactive forces [36]. Due to this, the proposed model that deals with the forces by coupling expressions of Rouse and size parameters unable to simulate the near-bed concentrations reasonably, as compared to the models presented in Ni et al. [19] and Zhong et al. [22] which used imported empirical functions from the respective experiments into their modelling. In addition, the boundary conditions for flow with transitional concentration (from dilute to dense concentrations) are hard to be fixed, and hence this difficulty may cause discrepancy in the proposed modelling (especially presented by results at Figure 15A,B).

Table 3. Data by Wang and Qian [36].

Test No.	d (mm)	ω_0 (cm/s)	u_* (cm/s)	$\overline{c}(\times 10^{-3})$
SF2	0.268	0.197	7.74	10.2
SF5	0.268	0.197	7.16	90.6
SM7	0.960	1.590	7.37	75.4
SC5	1.420	2.290	7.37	65.1

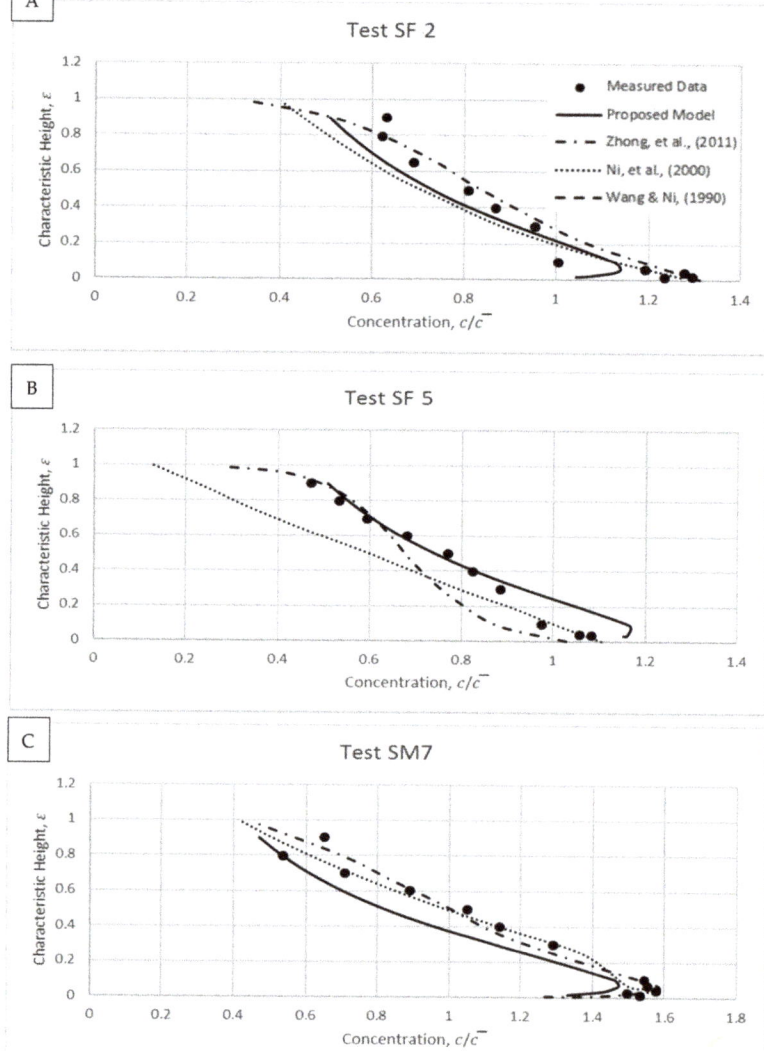

Figure 15. Cont.

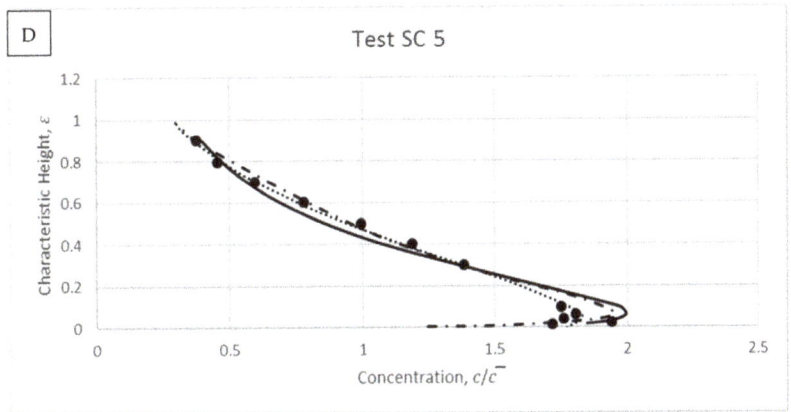

Figure 15. Modelled results and comparisons against experimental data of Wang and Qian [36].

4.3. Michalik

The experimental data of Michalik [35] quantified the sediment profile of hyper-concentrated flows. The sediment material used was sand with a mean diameter of 0.45 mm and with concentration ranging between $0.15 \leq \bar{c} \leq 0.54$ (all test conditions are shown in Table 4). In order to model Michalik's tests, this study uses the linear law within the proposed model. The test results are shown in Figure 16A–F, where the modelled results are compared to measurements. Ni et al. [19] and Zhong et al. [22] models are also incorporated into these figures to compare with the measurements and proposed model.

Table 4. Data by Michalik [35].

Test No.	d (mm)	ω_0 (cm/s)	u_* (cm/s)	$\bar{c} (\times 10^{-3})$
Run 1	0.45	6.15	15.56	150
Run 3	0.45	6.15	15.56	270
Run 4	0.45	6.15	15.56	310
Run 5	0.45	6.15	15.56	420
Run 6	0.45	6.15	15.56	450
Run 8	0.45	6.15	15.56	540

From the tested hyper-concentrated flows, the sediment concentration distribution illustrates a Type III profile. Within hyper-concentrated flow, the maximum in concentration is difficult to model accurately since there exists no distinct boundary where the sediment changes from bed to suspended load but rather the bed load can diffuse into the suspended state through a transitional region. As a result, the suspended load has sometimes been estimated as part of the bed load, which increases the discrepency in suspended solid modelling. To accurately define the transition region and consequently the location of this maximum turning point, a good estimation of ε_a and c_a is required [27]. In this study, the proposed model uses $\bar{c}$ for the mathematical modelling, since it seems acceptable to conclude that $\bar{c}$ is proportional to c_a [37] given that they are both invariant for a given experiment.

The proposed calculated results show reasonable agreement to the experimental data, in which it exhibits better fit to the measurements compared to the rest of models by Ni et al. [19] and Zhong et al. [22] in Figure 16A–F. The figures demonstrate a Type III profiles that fit into the hyper-concentrated distribution. Studies of hyper-concentrated flow show that the sediment concentration distribution becomes more homogenous when $\bar{c}$ increased as proven in Michalik [35]. Hence power law distributions might not be reasonable to model hyper-concentrated flow. Instead a linear model is adopted here and is evidenced to produce better accuracy. Even though the power-exponential models of Ni et al. [19] and Zhong et al. [22] are suitable to represent densely-concentrated flow, their accuracy are not encouraging compare to the measurements in majority of relatively hyper-concentrated $\bar{c}$ tests (i.e., at Figure 16D–F).

Additionally, one can observe that an increase in the mean concentration shifts the height of maximum concentration upwards through the characteristic height due to the increase of potential bed-load layer. As the mean concentration is increased, the near-bed region becomes more saturated. The tests in Figure 16 show that when $\bar{c} > 0.31$ the settling velocity has limited influence and the dominating interaction forces must arise from particle–particle reactions. In these hyper-concentrated tests of $\bar{c} > 0.31$, the proposed model represents the concentration distribution well. In particular, the proposed model accurately predicts the height at which the maximum concentration occurs for most cases.

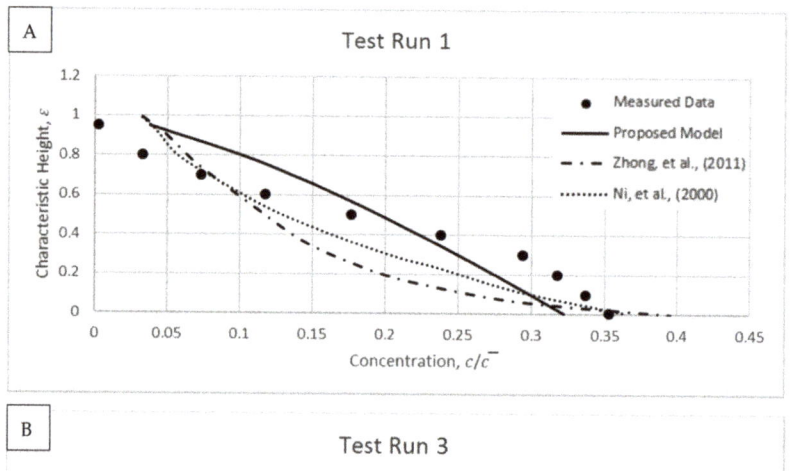

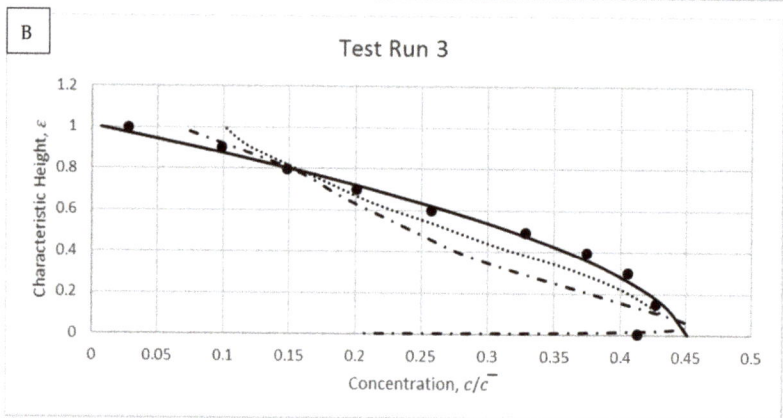

Figure 16. *Cont.*

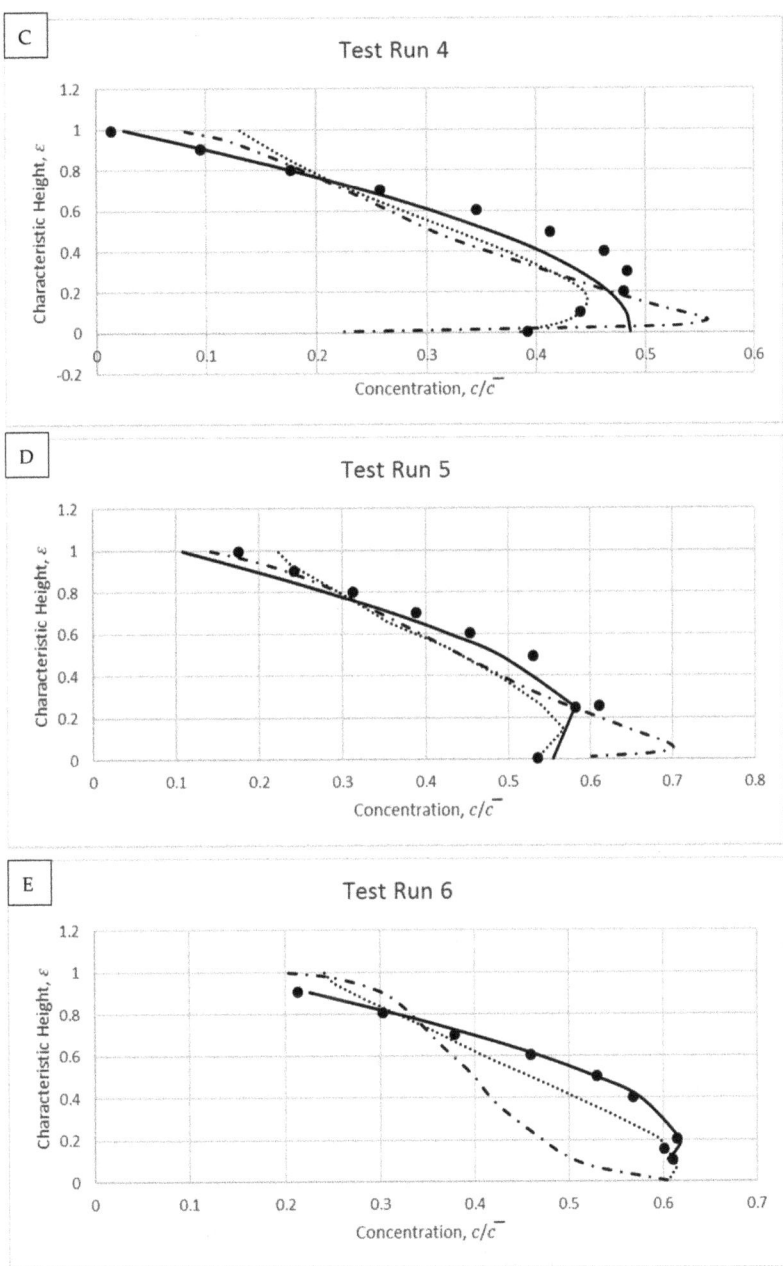

Figure 16. Cont.

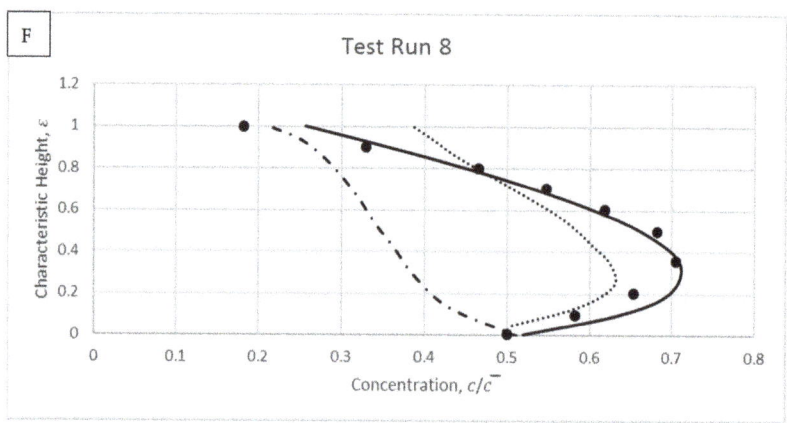

Figure 16. Showing modelled results and comparisons for experimental data of Michalik [35].

5. Conclusions

A parameterised power-linear coupled model has been introduced for inclusively computing the dilute- to hyper-concentrated distribution across the characteristic height within flow. The parameters used for the formulation of this model were size parameter, Rouse number, and mean concentration. As proven, the model is able to accurately compute the suspended sediment profile for a range of flow conditions including various Rouse numbers. The proposed model shows a reasonable accuracy for low and very high concentration tests across the Type I to III profiles. This can be seen from the comparisons with experimental data of Wang and Ni [31] on very dilute flows, Wang and Qian [36] on mixed dilute to dense flows, and Michalik [35] on hyper-concentrated flows. From the tests, the coupling approach of power to linear modelling has been proven to reasonably represent flow with a wide range of concentrations and sediment sizes.

This type of suspended modelling holds key importance to the accurate prediction of various natural flows, such as river, coastal, or flood flow. In flooded condition, the sediment mixture impacts the flow behaviour that can cause modelling failure in reproducing the real-world flood flow. With this analytical modelling study, the flood induced suspended sediment transport with wide range of dilute to dense concentration can be modelled adequately; and hence to provide the vital capability to flood flow modelling. Additionally, to further this work, different analytical modelling besides Rouse-based model can also be investigated.

Author Contributions: J.H.P.: writing—original draft preparation, writing—review and editing, funding acquisition, project administration, data curation, supervision; J.T.W.: writing—original draft preparation, writing—review and editing, data curation; M.A.K.: writing—review and editing, data curation; M.P.: writing—review and editing, data curation; H.P.: writing—review and editing, data curation; A.S.: writing—review and editing, data curation; P.R.H.: writing—review and editing, data curation; T.G.: writing—review and editing, supervision. All authors have read and agreed to the published version of the manuscript.

Funding: This research received no external funding.

Institutional Review Board Statement: Not applicable.

Informed Consent Statement: Not applicable.

Data Availability Statement: The data presented in this study are available on reasonable request from the corresponding author.

Conflicts of Interest: The authors declare no conflict of interest.

References

1. Pu, J.H.; Hussain, K.; Shao, S.-D.; Huang, Y.-F. Shallow sediment transport flow computation using time-varying sediment adaptation length. *Int. J. Sediment Res.* **2014**, *29*, 171–183. [CrossRef]
2. Pu, J.H.; Wei, J.; Huang, Y. Velocity Distribution and 3D Turbulence Characteristic Analysis for Flow over Water-Worked Rough Bed. *Water* **2017**, *9*, 668. [CrossRef]
3. Pu, J.H.; Huang, Y.; Shao, S.; Hussain, K. Three-Gorges Dam Fine Sediment Pollutant Transport: Turbulence SPH Model Simulation of Multi-Fluid Flows. *J. Appl. Fluid Mech.* **2016**, *9*, 1–10. [CrossRef]
4. Pu, J.H. Turbulent rectangular compound open channel flow study using multi-zonal approach. *Environ. Fluid Mech.* **2018**, *19*, 785–800. [CrossRef]
5. Pu, J.H.; Pandey, M.; Hanmaiahgari, P.R. Analytical modelling of sidewall turbulence effect on streamwise velocity profile using 2D approach: A comparison of rectangular and trapezoidal open channel flows. *HydroResearch* **2020**, *32*, 17–25. [CrossRef]
6. Rouse, H. Modern Conceptions of the Mechanics of Fluid Turbulence. *J. Pap. Trans. Am. Soc. Civ. Eng.* **1937**, *102*, 463–505.
7. Hsu, T.-J.; Jenkins, J.T.; Liu, P.L. On two-phase sediment transport: Dilute flow. *J. Geophys. Res. Space Phys.* **2003**, *108*, 3057. [CrossRef]
8. Huang, S.-H.; Sun, Z.-L.; Xu, D.; Xia, S.-S. Vertical distribution of sediment concentration. *J. Zhejiang Univ. A* **2008**, *9*, 1560–1566. [CrossRef]
9. Greimann, B.; Holly, F.M., Jr. Two-Phase Flow Analysis of Concentration Profiles. *J. Hydraul. Eng.* **2001**, *127*, 753–762. [CrossRef]
10. Jha, S.K.; Bombardelli, F.A. Two-phase modeling of turbulence in dilute sediment-laden, open-channel flows. *Environ. Fluid Mech.* **2009**, *9*, 237–266. [CrossRef]
11. Kundu, S.; Ghoshal, K. A mathematical model for type II profile of concentration distribution in turbulent flows. *Environ. Fluid Mech.* **2016**, *17*, 449–472. [CrossRef]
12. Pu, J.H.; Lim, S.Y. Efficient numerical computation and experimental study of temporally long equilibrium scour development around abutment. *Environ. Fluid Mech.* **2013**, *14*, 69–86. [CrossRef]
13. Liang, C.-F.; Abbasi, S.; Pourshahbaz, H.; Taghvaei, P.; Tfwala, S. Investigation of Flow, Erosion, and Sedimentation Pattern around Varied Groynes under Different Hydraulic and Geometric Conditions: A Numerical Study. *Water* **2019**, *11*, 235. [CrossRef]
14. Kundu, S.; Ghoshal, K. Explicit formulation for suspended concentration distribution with near-bed particle deficiency. *Powder Technol.* **2014**, *253*, 429–437. [CrossRef]
15. Wang, G.; Ni, J. The kinetic theory for dilute solid/liquid two-phase flow. *Int. J. Multiph. Flow* **1991**, *17*, 273–281. [CrossRef]
16. Goeree, J.C.; Keetels, G.H.; Munts, E.A.; Bugdayci, H.H.; Van Rhee, C. Concentration and velocity profiles of sediment-water mixtures using the drift flux model. *Can. J. Chem. Eng.* **2016**, *94*, 1048–1058. [CrossRef]
17. Pourshahbaz, H.; Abbasi, S.; Taghvaei, P. Numerical scour modeling around parallel spur dikes in FLOW-3D. *J. Drink. Water Eng. Sci.* **2017**, 1–16. [CrossRef]
18. Pourshahbaz, H.; Abbasi, S.; Pandey, M.; Pu, J.H.; Taghvaei, P.; Tofangdar, N. Morphology and hydrodynamics numerical simulation around groynes. *ISH J. Hydraul. Eng.* **2020**, 1–9. [CrossRef]
19. Ni, J.R.; Wang, G.Q.; Borthwick, A.G.L. Kinetic Theory for Particles in Dilute and Dense Solid-Liquid Flows. *J. Hydraul. Eng.* **2000**, *126*, 893–903. [CrossRef]
20. Van Rijn, L.C. Sediment Transport, Part II: Suspended Load Transport. *J. Hydraul. Eng.* **1984**, *110*, 1613–1641. [CrossRef]
21. McLean, S.R. On the calculation of suspended load for noncohesive sediments. *J. Geophys. Res. Space Phys.* **1992**, *97*, 5759. [CrossRef]
22. Zhong, D.; Wang, G.; Sun, Q. Transport Equation for Suspended Sediment Based on Two-Fluid Model of Solid/Liquid Two-Phase Flows. *J. Hydraul. Eng.* **2011**, *137*, 530–542. [CrossRef]
23. Fick, A. On liquid diffusion. *J. Membr. Sci.* **1995**, *100*, 33–38. [CrossRef]
24. Almedeij, J. Asymptotic matching with a case study from hydraulic engineering. *Proc. Recent Adv. Water Resour. Hydraul. Hydrol. Camb.* **2009**, 71–76. [CrossRef]
25. Einstein, H.A.; Qian, N. *Effects of Heavy Sediment Concentration Near the Bed on the Velocity and Sediment Distribution*; Omaha: U.S. Army Engineer Division, Missouri River: Omaha, NE, USA, 1955.
26. Bouvard, M.; Petković, S. Vertical dispersion of spherical, heavy particles in turbulent open channel flow. *J. Hydraul. Res.* **1985**, *23*, 5–20. [CrossRef]
27. Sumer, B.; Kozakiewicz, A.; Fredsoe, J.; Deigaard, R. Velocity and Concentration Profiles in Sheet-Flow Layer of Movable Bed. *J. Hydraul. Eng.* **1996**, *122*, 549–558. [CrossRef]
28. Kironoto, B.A.; Yulistiyanto, B. The validity of Rouse equation for predicting suspended sediment concentration profiles in transverse direction of uniform open channel flow. In Proceedings of the International Conference on Sustainable Development Water Waste Water Treatment, Yogyakarta, Indonesia, 14–15 December 2009.
29. Kumbhakar, M.; Ghoshal, K.; Singh, V.P. Derivation of Rouse equation for sediment concentration using Shannon entropy. *Phys. A Stat. Mech. Its Appl.* **2017**, *465*, 494–499. [CrossRef]
30. Coleman, N.L. Effects of Suspended Sediment on the Open-Channel Velocity Distribution. *Water Resour. Res.* **1986**, *22*, 1377–1384. [CrossRef]
31. Wang, G.; Ni, J. Kinetic Theory for Particle Concentration Distribution in Two-Phase Flow. *J. Eng. Mech.* **1990**, *116*, 2738–2748. [CrossRef]

32. Cellino, M.; Graf, W.H. Sediment-Laden Flow in Open-Channels under Noncapacity and Capacity Conditions. *J. Hydraul. Eng.* **1999**, *125*, 455–462. [CrossRef]
33. Muste, M.; Yu, K.; Fujita, I.; Ettema, R. Two-phase versus mixed-flow perspective on suspended sediment transport in turbulent channel flows. *Water Resour. Res.* **2005**, *41*, 41. [CrossRef]
34. Cheng, N.-S. Comparison of formulas for drag coefficient and settling velocity of spherical particles. *Powder Technol.* **2009**, *189*, 395–398. [CrossRef]
35. Michalik, M.A. Density patterns of the inhomogeneous liquids in the industrial pipe-lines measured by means of radiometric scanning. *La Houille Blanche* **1973**, 53–57. [CrossRef]
36. Wang, X.; Qian, N. Turbulence Characteristics of Sediment-Laden Flow. *J. Hydraul. Eng.* **1989**, *115*, 781–800. [CrossRef]
37. Ali, S.Z.; Dey, S. Mechanics of advection of suspended particles in turbulent flow. *Proc. R. Soc. A Math. Phys. Eng. Sci.* **2016**, *472*, 20160749. [CrossRef]

Article

Porosity Models for Large-Scale Urban Flood Modelling: A Review

Benjamin Dewals *[ID], Martin Bruwier, Michel Pirotton [ID], Sebastien Erpicum [ID] and Pierre Archambeau [ID]

Research unit Urban & Environmental Engineering (UEE), Hydraulics in Environmental and Civil Engineering (HECE), University of Liège, 4000 Liège, Belgium; martinbruwier@hotmail.com (M.B.); michel.pirotton@uliege.be (M.P.); s.erpicum@uliege.be (S.E.); pierre.archambeau@uliege.be (P.A.)
* Correspondence: b.dewals@uliege.be

Abstract: In the context of large-scale urban flood modeling, porosity shallow-water models enable a considerable speed-up in computations while preserving information on subgrid topography. Over the last two decades, major improvements have been brought to these models, but a single generally accepted model formulation has not yet been reached. Instead, existing models vary in many respects. Some studies define porosity parameters at the scale of the computational cells or cell interfaces, while others treat the urban area as a continuum and introduce statistically defined porosity parameters. The porosity parameters are considered either isotropic or anisotropic and depth-independent or depth-dependent. The underlying flow models are based either on the full shallow-water equations or approximations thereof, with various flow resistance parameterizations. Here, we provide a review of the spectrum of porosity models developed so far for large-scale urban flood modeling.

Keywords: urban flood modeling; porosity; shallow-water model

Citation: Dewals, B.; Bruwier, M.; Pirotton, M.; Erpicum, S.; Archambeau, P. Porosity Models for Large-Scale Urban Flood Modelling: A Review. *Water* **2021**, *13*, 960. https://doi.org/10.3390/w13070960

Academic Editors: James Shucksmith and Jorge Leandro

Received: 24 February 2021
Accepted: 26 March 2021
Published: 31 March 2021

Publisher's Note: MDPI stays neutral with regard to jurisdictional claims in published maps and institutional affiliations.

Copyright: © 2021 by the authors. Licensee MDPI, Basel, Switzerland. This article is an open access article distributed under the terms and conditions of the Creative Commons Attribution (CC BY) license (https://creativecommons.org/licenses/by/4.0/).

1. Introduction

Worldwide, climate evolution, population growth and rapid urbanization tend to increase urban flood risk [1,2]. Though this trend is well established, the magnitude of changes in flood risk and the distribution of risk in space and time remain highly uncertain [3]. Therefore, flood risk management should be guided by analyzing a high number of scenarios based on many runs of numerical models used for predicting flood hazard. This requires a high computational efficiency of the models, as it is also necessary for real-time forecasting of urban flooding, catchment-scale analyses, and interactive computations for risk communication [4]. Concurrently, high-resolution topographic data have become widely available. There is thus a need for high-performance urban flood models, which take benefit of available data to guide risk management and climate adaptation [5].

Meshing real-world urban areas for detailed flood modeling may prove very demanding. Indeed, a relatively fine discretization is required to capture relevant flow paths (voids in-between buildings) whose characteristic size is typically a few decameters, while computational domains covering urban areas may extend over hundreds of km². This makes fast computations particularly challenging [6]. Besides massive parallelization [4,7], another viable option for improving the computational efficiency of urban flood models consists of using subgrid modeling techniques, in which the computation is performed on a relatively coarse grid. At the same time, information on the sub-grid scale topography is preserved [8]. Porosity shallow-water models are a promising sub-grid modeling technique for large-scale urban flood modeling [9,10], as computation times two to three orders of magnitude smaller than standard shallow-water models were reported [11–13]. Over the last two decades, rapid advances have been made in the development of porosity shallow-water models (Figure 1).

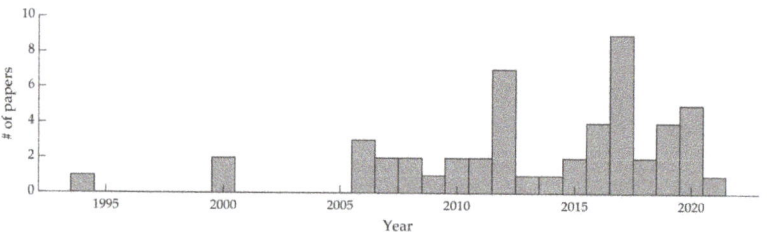

Figure 1. Number of papers included in this review as a function of the publication year.

Therefore, it is deemed timely to conduct a review of the many recent contributions in the field. Half of the studies included in this review were published over the last five years.

Porosity shallow-water models are based on a relatively coarse computational mesh, while so-called *porosity* parameters are introduced to account for topographic information available at a subgrid-scale [9,10]. This approach is similar to common practice in modeling flow in porous media, such as in groundwater modeling [14].

The presence of an obstacle, such as buildings, in an urban environment, has a threefold effect on the flow: they reduce the volume available for water storage; they channelize the flow along directional pathways defined by the arrangement of the obstacles; and they induce flow resistance due to various mechanisms, such as wakes [15]. The first effect is reproduced using a *storage porosity* parameter, which indicates the fraction of space available for mass and momentum storage. In all models, this storage porosity is consistently evaluated as the ratio of the volume of void in-between obstacles to the volume of a considered control volume. The other effects are accounted for in various ways, such as by means of additional porosity parameters characterizing the flow conveyance along with specific directions [9,11,15–17] or through directional flow resistance terms expressed in the tensor form [18,19].

Many flood models were developed based on the concept of porosity, but they vary greatly in terms of conceptual, mathematical and numerical formulations. As highlighted in Table 1,

- Porosity parameters are defined either as statistical descriptors of the urban area at large-scale [10,20] or from local geometric features [9,11,21];
- Models include either a single [10,20] or multiple porosity parameters [9,11,15–17];
- Effect of porosity in model fluxes and source terms is either isotropic [10,20] or anisotropic [9,11,12,15,16,21–23];
- Porosity parameters are either depth-independent [9–11,17] or depth-dependent [12,22,23];
- Models are expressed in differential [10,20,21] or in integral form [9,11];

Moreover, the underlying flow model may correspond to the complete shallow-water equations (dynamic wave) [9–12,15,20,22,23] or to an approximation thereof, such as the diffusive wave [24–26].

Table 1. Dualities in existing porosity models.

Porosity as a Statistical Descriptor [10,20]	Porosity as a Deterministic Geometric Parameter [9,11,21]
Single porosity parameter [10,20] (e.g., conveyance porosity equal to storage porosity)	Multiple porosity parameters [9,11,15–17]
Isotropic porosity effects [10,20]	Anisotropic porosity effects [9,11,12,15,16,21–23]
Depth-independent porosity [9–11,17]	Depth-dependent porosity [12,22,23]
Model expressed in differential form [10,20,21]	Model expressed in the integral form [9,11]
Numerical scheme limited to subcritical flow [16]	Shock-capturing schemes [9–12,15,20,22,23]
Shallow-water (dynamic wave) [9–12,15,20,22,23]	Diffusive wave approximation [24–26]
Isotropic flow resistance, e.g., [6]	Directional flow resistance [18,19]

Specific features are included in some models, such as separate flow paths within a single-cell thanks to a multilayered approach [24] or a multiple porosity model [17]. With this review, the authors aim to help the reader navigate through those various formulations of porosity shallow-water models for large-scale urban flood modeling.

In Section 2, we define the porosity parameters based on a control volume of relevance for urban flood modeling. As suggested by Table 1, there are multiple possibilities for classifying existing porosity shallow-water models. We opted for organizing the review in two-steps. Section 3 presents models in which porosity parameters are defined as statistical descriptors over an area sufficiently large to represent the urban area at a large-scale. In contrast, models that consider porosities defined based on local geometric parameters are detailed in Section 4. Figure 2 provides a schematic representation of the articulation between the major contributions to the field. They are all organized around a handful of landmark papers, as detailed in the following sections. Finally, Section 5 draws attention to recommended directions for future research.

2. Control Volume and Porosity Parameters

Almost all porosity shallow-water models aim at resolving the flow variables *on average* over a certain region of space. Hence, in the first place, these models are derived by integrating the flow governing equations over a control volume, as detailed in Appendix A of [10] or in [9]. A control volume of relevance for urban flood modeling is sketched in Figure 3. It is characterized by the presence of rigid obstacles (e.g., buildings) and water in-between. The control volume is delimited downward by the bottom elevation, upward by the water surface and laterally by vertical boundaries. The total volume of the control volume shown in Figure 3 is noted V, while V_f is the part of V filled with water (Figure 3a). Similarly, the contour of V is noted ∂V, while ∂V_f is the part of ∂V through which water can be exchanged (Figure 3b). The projection of the control volume on the horizontal plane $x-y$ is noted Ω, while, for a given arbitrary elevation z, the part Ω corresponding to voids is Ω_f (Figure 3c). The contour of Ω is noted $\partial \Omega$, while $\partial \Omega_f$ is the part of $\partial \Omega$ through which fluid can be exchanged (Figure 3d).

Note that depending on the particular type of porosity shallow-water model, the control volume shown in Figure 3 may correspond to a computational cell [9,11] or to a much wider area (e.g., a representative elementary volume, as discussed in Section 3).

Two types of porosities are used in porosity shallow-water models. First, the *storage porosity* ϕ, quantifies the volume of voids, i.e., the volume actually available to store water mass and momentum. For a given water level z in a control volume, the storage porosity is expressed mathematically as $\phi(z) = V_f(z)/V(z)$ [12,22,23]. As highlighted by [23], this definition of ϕ as a function of z is not univocal. It requires an assumption on the shape of the free surface. This shape is often assumed horizontal in the control volume [23].

When the obstacles may be considered prismatic for the range of water depths of interest and that none of them is submerged, the ratio of the volumes V_f and V become independent of the level z. Consequently, the following alternate definition was extensively used when referring to depth-independent storage porosity: $\phi = \Omega_f/\Omega$ [9–11,20].

The *conveyance porosity*, Ψ, quantifies the fraction of space available for mass and momentum exchange. Unlike the storage porosity, there is not a single clear-cut geometric definition Ψ. Depending on the models, the conveyance porosity is defined either statistically [10,20], or locally at the cell boundary ($\Psi(z) = \partial V_f(z)/\partial V(z)$, or $\Psi = \partial \Omega_f/\partial \Omega$ in the depth-independent case) [9,11], or at the level of a computational cell (and not just its boundary) [21]. The motivations for these various choices, as well as their implications on model accuracy and mesh sensitivity, are discussed in Section 4.

Note that care must be taken to the nomenclature, as the wording "storage" and "conveyance" porosity was not used uniformly across past studies (Table 2). Particularly, the conveyance porosity is also referred to as connectivity porosity, or areal porosity, among other terms.

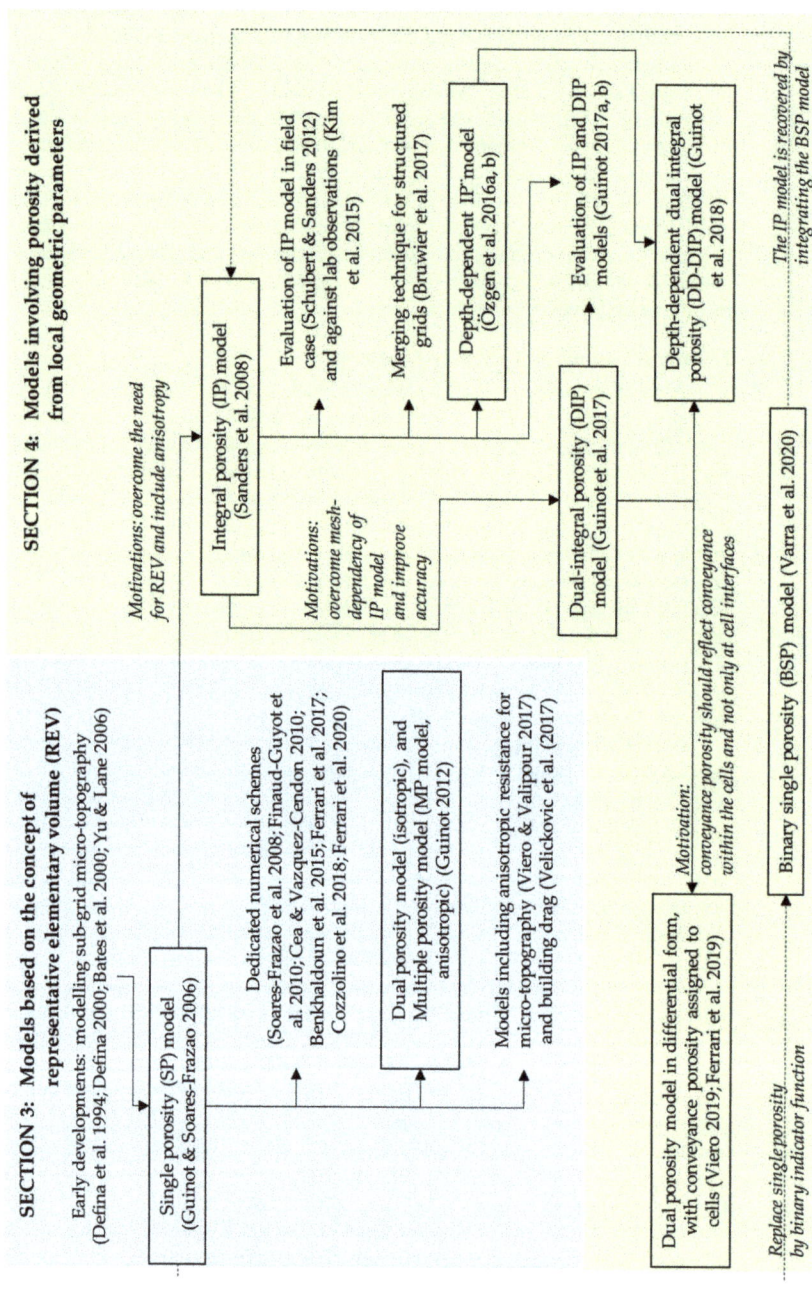

Figure 2. Articulation between selected major contributions to the development of porosity shallow-water models for large-scale urban flood modeling.

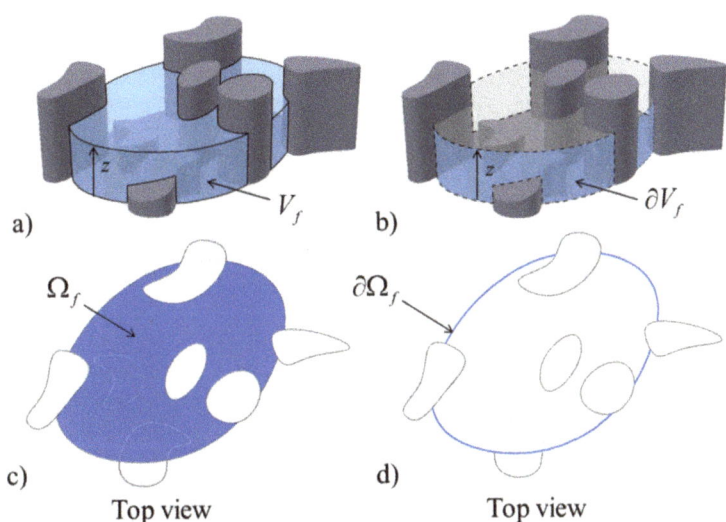

Figure 3. Sketch of a control volume and (**a**) the part V_f occupied by water, (**b**) the fluid–fluid exchange boundary ∂V_f, (**c**) a horizontal surface at a given level z and its part occupied by water Ω_f, as well as (**d**) the fluid–fluid exchange border $\partial \Omega_f$ of the horizontal surface.

Table 2. Nomenclature used in existing research to denote storage and conveyance porosities.

Context	References	Parameter Reflecting Storage Capacity in Control Volumes	Parameter Reflecting Fraction of Space Available for Flow Conveyance
Depth-independent porosity model	[1,2,6–8]	Storage porosity	Conveyance porosity
	[9–11]	Storage (or areal) porosity	Connectivity (or frontal) porosity
	[12,13]	1 − BCR, with BCR = building coverage ratio	1 − CRF, with CRF = conveyance reduction factors
Depth-dependent porosity model	[5,15]	Volumetric porosity	Areal porosity
	[16,17,20,21]		

3. Models Based on the Concept of Representative Elementary Volume

3.1. Representative Elementary Volume

Pioneering work on porosity shallow-water models was done by Defina et al. (1994) [27] to improve the numerical treatment of wetting and drying fronts when solving the full shallow-water equations. Their approach was later improved by Defina (2000) [28] and Bates (2000) [29]. In these models, the porosity was considered variable with the water depth because it was used as a statistical descriptor of sub-grid microtopographic features and not of buildings. Later, the concept of porosity was transposed to urban flood modeling as a building treatment method [10].

The reasoning underpinning the first porosity shallow-water models consists in idealizing the urban area as a fictitious continuum, in which the flow properties are described by statistical averages at a scale much larger than the scale of individual obstacles and pathways, but also sufficiently small compared to the extent of the whole urban area. This type of approach is widely used for modeling flow in porous media [14], such as in groundwater modeling, and it relies on the concept of *representative elementary volume* (REV). In general terms, the REV is defined as the smallest control volume for which the statistical properties of the porous medium become independent of the size of this control volume [14]. As pointed out by [17], in the case of a two-dimensional shallow flow model, the REV should normally be called representative elementary area [16], but the terminology REV is preserved here for the sake of consistency with most previous studies [9,17]. We consider these "two-dimensional REVs".

Provided that it exists in the considered medium, a REV can be defined around any arbitrary point (x,y), irrespective of the positioning of this specific point in water or in an obstacle (Figure 3). Indeed, in all cases, the porosity parameters can be evaluated as averages over the REV, which is much wider than the individual obstacles. It results that, in the REV, the mathematical expectation that a particular point is located in water or in a building is ϕ and $1 - \phi$, respectively.

3.2. Single Porosity Model

By phase-averaging the standard shallow-water equations over a REV containing fluid and obstacles, porosity shallow-water equations were derived by Guinot and Soares-Frazão (2006) [10] (Figure 2). They considered the porosity as depth-independent since it is used to represent the effect of buildings (assumed tall compared to the flow depth) and not of microtopographic features. In general, phase-averaging the shallow-water equations leads to two types of porosity parameters as previously defined: one expressing the available space for mass and momentum storage (ϕ), and the other one referring to the space available for mass and momentum exchange (Ψ). However, the authors of the first models of this type assumed that $\Psi = \phi$ [10]. This is the reason why this kind of model is called the *single-porosity* model (SP model).

In a perspective of space-averaging over a REV, a uniform porosity value was generally assigned to the computational cells in the urban area. This choice of a uniform porosity value makes the model unable to reproduce preferential flow directions resulting from directional pathways induced by the arrangement of the buildings at the subgrid-scale. The theoretical wave celerities are identical to those of the standard shallow-water equations [10,11].

The model of Guinot and Soares-Frazao (2006) is expressed in the differential form [10]. Indeed, flow variables (flow depth and depth-averaged velocities) associated with an arbitrary location (x,y) represent an average of the corresponding flow property over a control volume whose centroid is located at the coordinates (x,y). For sufficiently large control volumes (i.e., at minimum equal to the REV), the geometric properties and the flow variables averaged over these control volumes are continuous, differentiable and independent of the specific size chosen for the control volume. A practical advantage of this is that those models do not show an over-sensitivity to the design of the computational mesh. This also relates to the fact that a uniform porosity value was generally assigned to the computational cells in the urban area, no matter how much space is actually occupied by obstacles in each individual cell [18].

Note that, in real-world urban areas, a REV usually does not exist as it would extend beyond the limits of the urban area itself. Nevertheless, based on computational examples, Guinot (2012) [17] highlights that porosity approaches may nonetheless deliver results of practical relevance even for domain sizes smaller than that of the REV. The reason for this is that the errors arising from the porosity evaluation are commensurate with the degree of precision of other parameters or input data in shallow-water models [17].

The resolution of the SP model was performed with several numerical schemes based on the finite volume technique [30,31] and the use of various types of Riemann solvers [10,20,32–36]. Although the SP model was written in differential form, an integral form of the equations was solved numerically since finite volume schemes were used.

3.3. Introducing Anisotropy: Directional Drag, Multiple Porosity Model and Nonuniform Porosity

When the SP model is used with a uniform value of porosity throughout the urban area, a major limitation of such a modeling strategy is its inability to represent directionality [17]. The need for considering two different porosities (storage and conveyance) was pointed out by Lhomme (2006) [37]. However, although two distinct parameters were formally introduced in the governing equations, they were given equal values, and no insight was given on how to infer the value of the conveyance porosity from the geometry of the building footprints [37].

First attempts to introduce anisotropy in the SP model were made through the source term representing buildings drag. Indeed, two types of momentum losses are usually included in the SP model: those due to bottom and sidewall friction, as well as losses induced by the interplay occurring between the flow and the obstructions not explicitly resolved (e.g., wave reflections, building wakes . . .). In the first implementations of the SP model, the formulation of the corresponding additional sink term was isotropic, and the associated coefficients were taken equal along the x and y directions and evaluated by Borda-like formulations (for the case of a regular grid of buildings) [10,20] or through calibration [35]. In contrast, Velickovic et al. (2017) [18] represented directional effects by introducing a tensor of drag coefficients and amplification coefficients depending on the flow direction. The formulation was later questioned by Guinot (2017) [38], and the use of Borda-like formulae was also invalidated [17]. Similar to [18], a tensor form was used by [19] to model directional flow resistance in the case of overland flow and shallow inundation in agricultural landscapes.

Separately, Guinot (2012) [17] introduced anisotropy in a REV-based model by decomposing the domain into five types of regions: obstacles, regions with stagnant water, regions of isotropic 2D flow, several regions characterized by anisotropic 1D flow and interconnections between the 1D anisotropic flow regions. The immobile regions may be used to represent the wakes of buildings. The different regions exchange mass and momentum as a function of local differences in water levels. Hence, flow exchange coefficients, instead of drag coefficients, need to be calibrated. In each type of region, a specific formulation of the porosity shallow-water equations is used. This approach is called the *multiple porosity model* (MP model) and is reported to give more accurate results than the SP model [17]. Unlike in the SP model, the MP model's theoretical wave celerities differ from those in the standard shallow-water equations. The MP model of Guinot (2012) [17] may be reduced to an isotropic dual-porosity formulation if only obstacles, stagnant water and 2D isotropic flow regions are considered.

While most studies based on the SP model used a uniform value of porosity in the whole urban area, several authors [39–41] demonstrated the viability of finite volume schemes for the solution of SP models with a local storage porosity defined at the level of computational cells, regardless of the conceptual problems linked to the REV definition. Considering a porosity value variable from one cell to another in the SP model enables reproducing preferential flow paths and hence anisotropy. These models can be regarded as particular applications of the binary single porosity model recently proposed by Varra et al. (2020) [42] and described in Section 4.4.

4. Models Involving Porosity Derived from Local Geometric Parameters

4.1. Integral Porosity Model

Another line of research was also followed for the development of porosity shallow-water models. Along this line, the concept of REV is not used, and the equations are written in integral form for a control volume, which is either taken equal to a computational cell [9] or arbitrary (i.e., irrespective of a particular discretization) [11,23]. The main motivation for this approach was to overcome the theoretical limitation related to the inexistence of a REV in most real-world urban areas, as well as the inability of the SP model to reproduce directional effects.

In a landmark paper, Sanders et al. (2008) [9] used the Reynolds transport theorem and a binary density function to derive a macroscopic form of mass conservation and momentum equations, called the *integral porosity model* (IP model). In these equations, written in integral form, both storage and conveyance porosities are involved. The storage porosity ϕ is computed similarly as in the SP model but considering one computational cell instead of the REV as control volume. The conveyance porosity Ψ is defined as the fraction of a boundary of a computational cell, which contributes to mass and momentum exchange. Sanders et al. (2008) [9] indicate how to compute this parameter from geospatial data, such as classified aerial imagery or a digital elevation model and vector data describing building footprints.

Although the IP model was originally written in integral form, deriving a differential analog is useful for checking numerical convergence and for evaluating the wave celerities [17]. In a first attempt to do so, several authors estimated that the theoretical wave celerities of the IP model differ from those of the standard shallow-water equations [11,37,39]. In particular, they concluded that the conveyance porosity Ψ, which accounts for building obstruction to the flow, must be smaller than the storage porosity ϕ; otherwise, wave celerities larger than in the case without obstruction would be obtained, which appears unphysical [11,38]. Nonetheless, this constraint was recently questioned by Varra et al. (2020) [42], who proposed another differential equivalent of the IP model, as detailed in Section 4.4. According to [42], the differential analogs considered earlier cannot be used to evaluate wave celerities of the IP model.

Unlike the single porosity in the SP model, both the storage and the conveyance porosities introduced by [9] are defined locally at the level of computational cells or cell boundaries. Additionally, the conveyance porosity of [9] depends on the orientation of the cell boundary, which enables anisotropy of the urban area to be accounted for. These features make the model of [9] more accurate [13,39] and more suitable than the SP model for reproducing directional effects induced by obstructions not explicitly resolved in the computations, but it also makes this model overly sensitive to the mesh design [17,43]. Guidelines were formulated and assessed for designing suitable meshes.

A so-called *gap-conforming* mesh is recommended [9] to capture the anisotropy of urban networks through the conveyance porosity parameters by ensuring that computational edges intersect the obstacles indeed. However, these guidelines for mesh design do not completely fix the mesh over-sensitivity of the IP model and are not applicable to all types of meshes (e.g., Cartesian) [6,9,43]. A technique consisting of merging computational cells with low porosity values was proposed by [44] for the case of Cartesian grids.

To account for head losses induced by subgrid-scale buildings, the IP model uses a quadratic drag expression in the model equations [9]. It involves a drag coefficient and the projected area of the obstructions as seen by an observer moving along the flow direction. In general, these quantities are both direction- and flow-dependent, but a single scalar value was used by [9]. Determining these values for real-world applications is not straightforward. For the case of a field test with complex building geometry and topographic variations, a simplified version of the building drag term was tested by [6]: the flow-direction dependence of the frontal area of obstructions was ignored, and this quantity was computed as the average of the frontal area of obstructions over the directions of all cell boundaries.

Özgen et al. (2016) [12,22] proposed an extension of the IP model, in which the storage and conveyance porosities are depth-dependent. Approaches similar to that of Sanders et al. (2008) [9] and Özgen et al. (2016) [12] were adopted in multiple other studies, which account for obstruction-induced effects on storage and conveyance at the level of each computational cell [4,8,24–26,45–51]. Various approximations of the shallow-water equations were used in these studies (e.g., diffusive wave), and the storage and conveyance properties were generally considered as depth-dependent. For pluvial flooding applications, Chen et al. (2012) [25] derived an integral porosity model based on a diffusive wave approximation. No building drag term was considered. Representing separate flow paths within a single-cell of a Cartesian grid was made possible in an upgraded version of the same model [24]. This feature is based on a multilayered approach, and it shows similarity with the MP model [17].

4.2. Dual Integral Porosity Model

In another landmark paper, Guinot et al. (2017) [11] derived the *dual integral porosity model* (DIP model) considering an arbitrary control volume (not necessarily linked to a computational cell). The DIP model is an extension of the IP model, which aims at correcting discrepancies between wave celerities obtained from the IP model and from refined calculations based on the standard shallow-water equations. Compared to the IP model, the DIP model contains three major conceptual improvements: (i) porosity and flow variables are defined separately for control volumes and boundaries, and a closure scheme is proposed to link control volume-based and boundary-based quantities; (ii) a new transient momentum dissipation mechanism active for positive waves is introduced, and (iii) an anisotropic drag force model is formulated.

As shown by [17], when a positive wave propagates in an urban area, wave reflections occur against the buildings and generate moving bores. The forces exerted by the building walls are opposed to the average flow velocity and thus contribute to dissipating momentum. Similar bores do not occur in the case of steady flow nor for decreasing water levels [17]. This momentum dissipation mechanism cannot be described by an equation of state, i.e., involving only the flow variables. Hence, it cannot be reproduced by means of a building drag term [23]. Therefore, Guinot et al. (2017) [11] introduced this dissipation mechanism directly in the fluxes of the model by means of a tensor whose elements need to be calibrated. It is active only under transient conditions involving positive waves (rising water levels).

Note that the closure scheme proposed by Guinot et al. (2017) [11] leads to using only the storage porosity in the continuity fluxes. This contrasts with the original IP model, but it makes the new continuity equation consistent with that originally derived by Defina [28]. For the DIP model to be well-posed, it is necessary that the conveyance porosity is lower than the storage porosity, as mentioned above for the IP model [38].

Based on a set of 96 benchmarks, enabling direct validation of flux closures and source terms, the DIP model was shown to outperform the IP and SP models [38]. The DIP model was also shown to be substantially less sensitive to mesh design than the IP model [43]. Similar to the extension brought by [12] to the IP model, Guinot et al. (2018) [23] proposed a new formulation of the DIP model, in which the porosities are depth-dependent, and the model is adapted to handle submerged obstructions. In this study, the superiority of the DIP model over the IP model is confirmed, and the transient momentum dissipation mechanism is shown to be essential [23].

4.3. Alternate Uses and Definitions of Conveyance Porosities

To further reduce the model sensitivity to the design of the mesh, Viero (2019) [16] implemented a dual-porosity model, in which the conveyance porosity is not evaluated locally at the cell *interfaces* but at the level of each computational *cell*, and it is defined along mutually orthogonal principal directions. It is assumed that water flows through the narrowest cross-section over the computational cell, as already assumed by [44] among others. This is justified by the occurrence of most dissipation at locations where velocity is the largest and by the fact that the effective length of the narrowest section is longer than its geometric length due to the jet developing downstream of a contraction [16]. Unlike previous implementations of the IP and DIP models using collocated finite volume schemes, a finite element scheme was used on a staggering unstructured mesh [16]. Tests against refined numerical solutions and experimental data suggest that the model sensitivity to the mesh design is acceptable in the tested configurations.

With the same objective of further reducing model mesh sensitivity, Ferrari et al. (2019) [15] considered an isotropic single porosity formulation for the fluxes and used anisotropic conveyance porosities to estimate flow resistance by means of a directionally dependent tensor formulation. Like in [16], the effective velocity for evaluating losses is determined from the narrowest cross-section in the considered urban district. The authors conclude that the model is not oversensitive to mesh resolution and design, but they call for more research on the determination of the conveyance porosity for real-world urban areas. This issue has been addressed by Ferrari and Viero (2020) [21], who detail an algorithm for computing distributed cell-based conveyance porosities as needed in the dual-porosity models. It is based on the analysis of the footprints of buildings and obstacles on Cartesian grids and uses mutually orthogonal principal directions. This approach performs well in the presence of a single dominant obstacle in the cell but not with multiple obstacles. Therefore, more research is needed regarding the modeling of conveyance porosity.

4.4. Binary Single Porosity Model

In a recent theoretical contribution, Varra et al. (2020) [42] introduced the *binary single porosity model* (BSP model), a novel local, differential porosity model formulation derived regardless of the existence of a REV. The BSP model adopts the same mathematical formulation as the SP model, the REV-based porosity parameter of the SP model being replaced by a binary indicator (equal to unity in the water and to zero in the obstacles). Therefore, derivatives in the BSP differential form must be understood in the sense of generalized functions (distributions). The IP model may be recovered from the BSP model by integration in space. As the derivation of the BSP model does not involve space averaging, the flow variables are pointwise and not space averaged values.

Varra et al. (2020) [42] claim that a suitable Riemann solver has the potential to take into account energy loss due to wave reflections, hence reducing the need to resort to additional drag terms. In addition, additional stationary dissipation is needed through porosity reductions in supercritical flow.

Despite encouraging results obtained so far with the dual porosity models [11,16,23,38,43], Varra et al. (2020) [42] indicates that the use of different storage and conveyance porosities in the mathematical model formulation in differential form violates the Galilean invariance and that the difference between storage and conveyance porosities arises in the numerical discretization, but should not be introduced in the model mathematical formulation. This further emphasizes the need for additional research on the modeling of anisotropic conveyance effects in porosity shallow-water models.

5. Directions for Further Research

This paper reviews the various porosity shallow-water models developed so far for large-scale urban flood modeling. Two main families of porosity models can be distinguished depending on the scale at which porosity parameters are determined (REV-based porosity vs. porosity derived from local geometric data). Recent developments have been numerous. They have addressed multiple aspects of the models, such as more physically grounded modeling of momentum dissipation mechanisms, enhanced determination of conveyance porosities, strategies to mitigate mesh over-sensitivity of integral porosity models or new insights into the theoretical formulation of the models.

Despite many efforts devoted to the formulation of models for building drag and other dissipation mechanisms, none of the current models is complete [38]. There are still gaps in knowledge regarding not only the calibration but also the structure of dissipation mechanism models adapted to porosity shallow-water equations for large-scale urban flood modeling. This calls for more research on both the conceptual and numerical aspects [42].

Recent advances in porosity models were not all evaluated based on the same test cases. This may influence conclusions drawn on the model's performance, such as accuracy or degree of mesh sensitivity. The scientific community would highly benefit from the setup of a series of accepted benchmarks against which every new contribution could be assessed. This would take the form of an evolving, shared database of test cases as it does exist in other fields. Such test cases should incorporate a blend of idealized, synthetic [52] and fully realistic configurations, including high-quality field observations of flow depth and velocity. Particularly valuable are direct evaluations of flux and source terms [38], as well as disentangling structural, scaling and porosity model errors [13].

The transfer of porosity shallow-water models from research to practice poses specific challenges [4]. Guidelines should be developed to enable practitioners to achieve optimal mesh design and model calibration. However, a general methodology for model parametrization for real-world urban areas remains a research question. Strategies could be elaborated for calibrating porosity models using fine-scale reference model runs over only a limited domain or for optimally combining porosity and detailed models using domain decomposition or nested models.

Flow modeling results are often used as input for complementary analyses, such as damage modeling, solute [46] or sediment transport and morphodynamic modeling. It is, therefore, necessary to assess whether the porosity models succeed in predicting, at the right scale, the flow variables needed for these complementary analyses and which postprocessing steps may be necessary [53].

Author Contributions: Conceptualization, B.D., P.A. and M.B.; formal analysis, B.D. and M.B.; writing—original draft, B.D. and M.B.; writing—review and editing, B.D., P.A., S.E., M.P. All authors have read and agreed to the published version of the manuscript.

Funding: This research received no external funding.

Informed Consent Statement: Not applicable.

Acknowledgments: The authors gratefully acknowledge two anonymous reviewers whose insightful comments substantially improved several aspects of the manuscript.

Conflicts of Interest: The authors declare no conflict of interest.

References

1. Ward, P.J.; Blauhut, V.; Bloemendaal, N.; Daniell, E.J.; De Ruiter, C.M.; Duncan, J.M.; Emberson, R.; Jenkins, F.S.; Kirschbaum, D.; Kunz, M.; et al. Review article: Natural hazard risk assessments at the global scale. *Nat. Hazards Earth Syst. Sci.* **2020**, *20*, 1069–1096. [CrossRef]
2. Aerts, J.C.J.H.; Botzen, W.J.W.; Emanuel, K.; Lin, N.; De Moel, H.; Michel-Kerjan, E.O. Climate adaptation: Evaluating flood resilience strategies for coastal megacities. *Science* **2014**, *344*, 473–475. [CrossRef]
3. IPCC. *Managing the Risks of Extreme Events and Disasters to Advance Climate Change Adaptation—A Special Report of Working Groups I and II of the IPCC*; Field, C.B., Barros, V., Stocker, T.F., Qin, D., Dokken, D.J., Ebi, K.L., Mastrandrea, M.D., Mach, K.J., Plattner, G.-K., Allen, S.K., et al., Eds.; Cambridge University Press: Cambridge, UK, 2012.
4. Sanders, B.F.; Schubert, J.E. PRIMo: Parallel raster inundation model. *Adv. Water Resour.* **2019**, *126*, 79–95. [CrossRef]
5. Dottori, F.; Di Baldassarre, G.; Todini, E. Detailed data is welcome, but with a pinch of salt: Accuracy, precision, and uncertainty in flood inundation modeling. *Water Resour. Res.* **2013**, *49*, 6079–6085. [CrossRef]
6. Schubert, J.E.; Sanders, B.F. Building treatments for urban flood inundation models and implications for predictive skill and modeling efficiency. *Adv. Water Resour.* **2012**, *41*, 49–64. [CrossRef]
7. Yu, D. Parallelization of a two-dimensional flood inundation model based on domain decomposition. *Environ. Model. Softw.* **2010**, *25*, 935–945. [CrossRef]
8. McMillan, H.K.; Brasington, J. Reduced complexity strategies for modelling urban floodplain inundation. *Geomorphology* **2007**, *90*, 226–243. [CrossRef]
9. Sanders, B.F.; Schubert, J.E.; Gallegos, H.A. Integral formulation of shallow-water equations with anisotropic porosity for urban flood modeling. *J. Hydrol.* **2008**, *362*, 19–38. [CrossRef]
10. Guinot, V.; Soares-Frazão, S. Flux and source term discretization in two-dimensional shallow water models with porosity on unstructured grids. *Int. J. Numer. Methods Fluids* **2006**, *50*, 309–345. [CrossRef]
11. Guinot, V.; Sanders, B.F.; Schubert, J.E. Dual integral porosity shallow water model for urban flood modelling. *Adv. Water Resour.* **2017**, *103*, 16–31. [CrossRef]
12. Özgen, I.; Liang, D.; Hinkelmann, R. Shallow water equations with depth-dependent anisotropic porosity for subgrid-scale topography. *Appl. Math. Model.* **2016**, *40*, 7447–7473. [CrossRef]
13. Kim, B.; Sanders, B.F.; Famiglietti, J.S.; Guinot, V. Urban flood modeling with porous shallow-water equations: A case study of model errors in the presence of anisotropic porosity. *J. Hydrol.* **2015**, *523*, 680–692. [CrossRef]
14. Bear, J. *Dynamics of Fluids in Porous Media*; Dover Publications Inc.: New York, NY, USA, 1988.
15. Ferrari, A.; Viero, D.P.; Vacondio, R.; Defina, A.; Mignosa, P. Flood inundation modeling in urbanized areas: A mesh-independent porosity approach with anisotropic friction. *Adv. Water Resour.* **2019**, *125*, 98–113. [CrossRef]
16. Viero, D.P. Modelling urban floods using a finite element staggered scheme with an anisotropic dual porosity model. *J. Hydrol.* **2019**, *568*, 247–259. [CrossRef]

17. Guinot, V. Multiple porosity shallow water models for macroscopic modelling of urban floods. *Adv. Water Resour.* **2012**, *37*, 40–72. [CrossRef]
18. Velickovic, M.; Zech, Y.; Soares-Frazão, S. Steady-flow experiments in urban areas and anisotropic porosity model. *J. Hydraul. Res.* **2017**, *55*, 85–100. [CrossRef]
19. Viero, D.P.; Valipour, M. Modeling anisotropy in free-surface overland and shallow inundation flows. *Adv. Water Resour.* **2017**, *104*, 1–14. [CrossRef]
20. Soares-Frazão, S.; Lhomme, J.; Guinot, V.; Zech, Y. Two-dimensional shallow-water model with porosity for urban flood modelling. *J. Hydraul. Res.* **2008**, *46*, 45–64. [CrossRef]
21. Ferrari, A.; Viero, D.P. Floodwater pathways in urban areas: A method to compute porosity fields for anisotropic subgrid models in differential form. *J. Hydrol.* **2020**, *589*. [CrossRef]
22. Özgen, I.; Zhao, J.; Liang, D.; Hinkelmann, R. Urban flood modeling using shallow water equations with depth-dependent anisotropic porosity. *J. Hydrol.* **2016**, *541*, 1165–1184. [CrossRef]
23. Guinot, V.; Delenne, C.; Rousseau, A.; Boutron, O. Flux closures and source term models for shallow water models with depth-dependent integral porosity. *Adv. Water Resour.* **2018**, *122*, 1–26. [CrossRef]
24. Chen, A.S.; Evans, B.; Djordjević, S.; Savić, D.A. Multi-layered coarse grid modelling in 2D urban flood simulations. *J. Hydrol.* **2012**, *470–471*, 1–11. [CrossRef]
25. Chen, A.S.; Evans, B.; Djordjević, S.; Savić, D.A. A coarse-grid approach to representing building blockage effects in 2D urban flood modelling. *J. Hydrol.* **2012**, *426–427*, 1–16. [CrossRef]
26. Yu, D.; Lane, S.N. Urban fluvial flood modelling using a two-dimensional diffusion-wave treatment, part 2: Development of a sub-grid-scale treatment. *Hydrol. Process.* **2006**, *20*, 1567–1583. [CrossRef]
27. Defina, A.; D'Alpaos, L.; Matticchio, B. New set of equations for very shallow water and partially dry areas suitable to 2D numerical models. In *Proceedings of the Specialty Conference on Modelling of Flood Propagation Over Initially Dry Areas, Milan, Italy, 29 June–1 July 1994*; ASCE: Reston, VA, USA, 1994; pp. 72–81.
28. Defina, A. Two-dimensional shallow flow equations for partially dry areas. *Water Resour. Res.* **2000**, *36*, 3251–3264. [CrossRef]
29. Bates, P.D. Development and testing of a subgrid-scale model for moving-boundary hydrodynamic problems in shallow water. *Hydrol. Process.* **2000**, *14*, 2073–2088. [CrossRef]
30. Cozzolino, L.; Pepe, V.; Cimorelli, L.; D'Aniello, A.; Della Morte, R.; Pianese, D. The solution of the dam-break problem in the porous shallow water equations. *Adv. Water Resour.* **2018**, *114*, 83–101. [CrossRef]
31. Mohamed, K. A finite volume method for numerical simulation of shallow water models with porosity. *Comput. Fluids* **2014**, *104*, 9–19. [CrossRef]
32. Ferrari, A.; Vacondio, R.; Mignosa, P. A second-order numerical scheme for the porous shallow water equations based on a DOT ADER augmented Riemann solver. *Adv. Water Resour.* **2020**, *140*. [CrossRef]
33. Ferrari, A.; Vacondio, R.; Dazzi, S.; Mignosa, P. A 1D–2D shallow water equations solver for discontinuous porosity field based on a generalized Riemann problem. *Adv. Water Resour.* **2017**, *107*, 233–249. [CrossRef]
34. Benkhaldoun, F.; Elmahi, I.; Moumna, A.; Seaid, M. A non-homogeneous Riemann solver for shallow water equations in porous media. *Appl. Anal.* **2016**, *95*, 2181–2202. [CrossRef]
35. Cea, L.; Vázquez-Cendón, M.E. Unstructured finite volume discretization of two-dimensional depth-averaged shallow water equations with porosity. *Int. J. Numer. Methods Fluids* **2010**, *63*, 903–930. [CrossRef]
36. Finaud-Guyot, P.; Delenne, C.; Lhomme, J.; Guinot, V.; Llovel, C. An approximate-state Riemann solver for the two-dimensional shallow water equations with porosity. *Int. J. Numer. Methods Fluids* **2010**, *62*, 1299–1331. [CrossRef]
37. Lhomme, J. One-Dimensional, Two-Dimensional and Macroscopic Approaches to Urban Flood Modelling. Ph.D.Thesis, Montpellier 2 University, Montpellier, France, 2006.
38. Guinot, V. A critical assessment of flux and source term closures in shallow water models with porosity for urban flood simulations. *Adv. Water Resour.* **2017**, *109*, 133–157. [CrossRef]
39. Özgen, I.; Zhao, J.-H.; Liang, D.-F.; Hinkelmann, R. Wave propagation speeds and source term influences in single and integral porosity shallow water equations. *Water Sci. Eng.* **2017**, *10*, 275–286. [CrossRef]
40. Velickovic, M.; Van Emelen, S.; Zech, Y.; Soares-Frazão, S. Shallow-water model with porosity: Sensitivity analysis to head losses and porosity distribution. In *Proceedings of the River flow 2010: International Conference on Fluvial Hydraulics, Braunschweig, Germany, 8–10 September 2010*; Dittrich, A., Koll, K., Aberle, J., Geisenhainer, P., Eds.; Bundesanstalt für Wasserbau: Karlsruhe, Germany, 2010; Volume 2, pp. 613–620.
41. Soares-Franzão, S.; Franzini, F.; Linkens, J.; Snaps, J.-C. Investigation of distributed-porosity fields for urban flood modelling using single-porosity models. In Proceedings of the E3S Web of Conferences, Lyon, France, 6–8 April 2018; Volume 40, p. 06040.
42. Varra, G.; Pepe, V.; Cimorelli, L.; Della Morte, R.; Cozzolino, L. On integral and differential porosity models for urban flooding simulation. *Adv. Water Resour.* **2020**, *136*. [CrossRef]
43. Guinot, V. Consistency and bicharacteristic analysis of integral porosity shallow water models. Explaining model oversensitivity to mesh design. *Adv. Water Resour.* **2017**, *107*, 43–55. [CrossRef]
44. Bruwier, M.; Archambeau, P.; Erpicum, S.; Pirotton, M.; Dewals, B. Shallow-water models with anisotropic porosity and merging for flood modelling on Cartesian grids. *J. Hydrol.* **2017**, *554*, 693–709. [CrossRef]

45. Li, Z.; Hodges, B.R. On modeling subgrid-scale macro-structures in narrow twisted channels. *Adv. Water Resour.* **2020**, *135*. [CrossRef]
46. Li, Z.; Hodges, B.R. Modeling subgrid-scale topographic effects on shallow marsh hydrodynamics and salinity transport. *Adv. Water Resour.* **2019**, *129*, 1–15. [CrossRef]
47. Shamkhalchian, A.; De Almeida, G.A.M. Upscaling the shallow water equations for fast flood modelling. *J. Hydraul. Res.* **2020**. [CrossRef]
48. Wu, G.; Shi, F.; Kirby, J.T.; Mieras, R.; Liang, B.; Li, H.; Shi, J. A pre-storage, subgrid model for simulating flooding and draining processes in salt marshes. *Coast. Eng.* **2016**, *108*, 65–78. [CrossRef]
49. Volp, N.D.; Van Prooijen, B.C.; Stelling, G.S. A finite volume approach for shallow water flow accounting for high-resolution bathymetry and roughness data. *Water Resour. Res.* **2013**, *49*, 4126–4135. [CrossRef]
50. Neal, J.; Schumann, G.; Bates, P. A subgrid channel model for simulating river hydraulics and floodplain inundation over large and data sparse areas. *Water Resour. Res.* **2012**, *48*, W11506. [CrossRef]
51. Casulli, V.; Stelling, G.S. Semi-implicit subgrid modelling of three-dimensional free-surface flows. *Int. J. Numer. Methods Fluids* **2011**, *67*, 441–449. [CrossRef]
52. Bruwier, M.; Mustafa, A.; Aliaga, D.G.; Archambeau, P.; Erpicum, S.; Nishida, G.; Zhang, X.; Pirotton, M.; Teller, J.; Dewals, B. Influence of urban pattern on inundation flow in floodplains of lowland rivers. *Sci. Total Environ.* **2018**, *622–623*, 446–458. [CrossRef]
53. Carreau, J.; Guinot, V. A PCA spatial pattern based artificial neural network downscaling model for urban flood hazard assessment. *Adv. Water Resour.* **2021**, *147*. [CrossRef]

Article

Development of a Simulation Model for Real-Time Urban Floods Warning: A Case Study at Sukhumvit Area, Bangkok, Thailand

Detchphol Chitwatkulsiri [1,*], Hitoshi Miyamoto [1] and Sutat Weesakul [2]

1 Department of Civil Engineering, Shibaura Institute of Technology, 3-7-5 Toyosu, Koto-ku, Tokyo 135-8548, Japan; miyamo@shibaura-it.ac.jp
2 Water Engineering and Management, School of Engineering and Technology, Asian Institute of Technology, Klong Luang 12120, Pathum Thani, Thailand; sutat@ait.ac.th
* Correspondence: na20109@shibaura-it.ac.jp

Abstract: Increasingly frequent, high-intensity rain events associated with climatic change are driving urban drainage systems to function beyond their design discharge capacity. It has become an urgent issue to mitigate the water resource management challenge. To address this problem, a real-time procedure for predicting the inundation risk in an urban drainage system was developed. The real-time procedure consists of three components: (i) the acquisition and forecast of rainfall data; (ii) rainfall-runoff modeling; and (iii) flood inundation mapping. This real-time procedure was applied to a drainage system in the Sukhumvit area of Bangkok, Thailand, to evaluate its prediction efficacy. The results showed precisely that the present real-time procedure had high predictability in terms of both the water level and flood inundation area mapping. It could also determine hazardous areas with a certain amount of lead time in the drainage system of the Sukhumvit area within an hour of rainfall data. These results show the real-time procedure could provide accurate flood risk warning, resulting in more time to implement flood management measures such as pumping and water gate operations, or evacuation.

Keywords: urban flood management; flood forecasting; weather radar; integrated hydraulic modeling; and evacuation lead time

Citation: Chitwatkulsiri, D.; Miyamoto, H.; Weesakul, S. Development of a Simulation Model for Real-Time Urban Floods Warning: A Case Study at Sukhumvit Area, Bangkok, Thailand. Water 2021, 13, 1458. https://doi.org/10.3390/w13111458

Academic Editor: Marco Franchini

Received: 8 April 2021
Accepted: 19 May 2021
Published: 22 May 2021

Publisher's Note: MDPI stays neutral with regard to jurisdictional claims in published maps and institutional affiliations.

Copyright: © 2021 by the authors. Licensee MDPI, Basel, Switzerland. This article is an open access article distributed under the terms and conditions of the Creative Commons Attribution (CC BY) license (https://creativecommons.org/licenses/by/4.0/).

1. Introduction

Globally, flooding continues to be a challenge for urban areas, a problem increasingly exacerbated by higher frequency, intense rainfall events [1–3]. Although there is a trend towards implementing water-sensitive urban design (WSUD) or nature-based solutions (NBS), in conjunction with hard engineering approaches [4–9] system wide development and improvement of urban drainage systems continue to be a financial and planning challenge for many municipalities [10–13]. A complementary approach to physical improvements of the drainage system that have the potential to aid in reducing flood damage and loss of life is the development of a real-time flood warning system. Research has been conducted with respect to real-time urban flood warning systems that include the combination of hydraulic and probabilistic modeling for real-time urban flood prediction [14], the application of recurrent neural networks for urban flood control in Taiwan [15], the use of radar images for urban flood detection in the UK [16], and an analysis sensitivity to spatiotemporal resolution of rainfall input and hydro-logic modeling for flash flood forecasting in USA [17], but demonstration of the real-time forecasting techniques remains limited.

Bangkok, Thailand, regularly experiences both large-scale fluvial and localized pluvial flooding that results in considerable damage and that is exacerbated by a number of factors, including rapid urbanization, a change of upstream condition in the northern part of Thailand, limitation of drainage capacity design, tidal effect, land subsidence from groundwater consumption and climate change [18–22]. The Sukhumvit area of downtown Bangkok is a vibrant and important commercial, tourist, and residential hub and, therefore, was chosen as our study site. The objective of the study, then, was to develop and demonstrate a real-time flood forecasting procedure that linked existing ground-level rain gauges, weather radar, a dynamic rainfall forecasting package and a conceptual, deterministic hydrologic/hydraulic model to simulate the drainage system responses in Sukhumvit to sudden heavy rainfall. The goal of the real-time procedure was to provide flood risk warnings sooner than currently is possible and allow earlier implementation of flood management measures such as pumping and water gate operations, or evacuation.

2. Materials and Methods

2.1. Study Area

Bangkok experiences a Tropical Savanna type (Aw) climate, with distinct rainy and dry seasons. Mean annual precipitation is 1651 mm, with about 85% falling in the rainy season, May to October. The Sukhumvit area of Bangkok, located north of the Chao Phraya River, is densely urbanized and one of the central business districts (Figure 1). The study area bounded by canals and the Chao Phraya River, is 24 sq.km, with a population density of 8400 persons/sq.km. The study catchment is quite flat, with average elevation being around 0.4 to 1.0 m above mean sea level (MSL), while maximum levee elevation along Klong (Canal) Sean Seap and Klong Tan/Phra Khanong are 1.2 m above MSL [19].

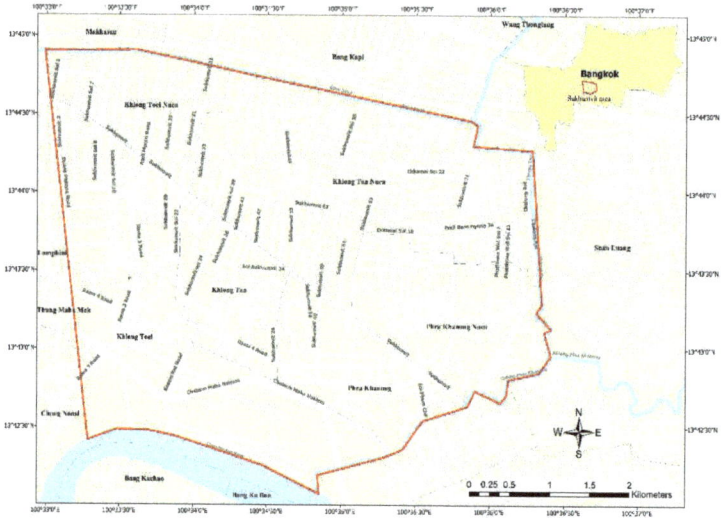

Figure 1. Map of Sukhumvit area, Bangkok, Thailand. The boundary of this study area is from one of the water management polder areas in Bangkok [19].

The Sukhumvit area has a primary drainage system consisting of surrounding canals and the secondary drainage system inside the area. Because of the relatively flat topography of the area, stormwater drainage via gravity flow can be a challenge and pumping stations, therefore, are an integral component of the drainage system. When the water level in the primary drainage system around Sukhumvit area increases, drainage of the secondary system becomes a particular problem. The study reported here is built on some previous modelling efforts for the Sukhumvit area. Weesakul et al. [23] used a street network connected to a pipe flow model and developed inundation maps using MIKE 11 GIS software which is a river modelling system that has been developed by the Danish Hydraulic Institute (DHI) [24]. The representation of the flood was made in a digital elevation model (DEM) layer. Nguyen [25] studied the rainfall forecasting and real-time hydrologic information system for Bangkok with an urban drainage model application for Sukhumvit. This work used a 1D modelling approach for pipe flow and the street flow network. The streets were created as channel flow elements with cross sections similar to the open canals that are connected to the pipe flow network for calculation of surcharged or flooded water volume. Shrestha [10] studied the impact of climate change on urban flooding by applying different climate change precipitation scenarios and simulated runoff by coupling a 1D and 2D flood model in this area. This study mainly aimed to develop a real-time procedure for predicting the inundation risk in this area by using a combination of recorded and forecasted rainfall data to simulated runoff and to provide an early warning for flood location.

2.2. The Framework of the Real-Time Urban Drainage Model

The development of a real-time urban drainage procedure and model for the Sukhumvit area was the main objective of this research. The procedure consists of three com-ponents as summarized in Figure 2. The first component was to obtain real-time, near-real time, and historical rainfall data from two different sources, a network of on-ground rain gauge stations in the general study area and the Bangkok Metropolitan Administration Department of Drainage and Sewerage (BMA, DDS) C-Band Radar at the Nong Jok Radar Station. Subsequently, forecasting of rainfall moving into the study area was undertaken using a translation method. This combination of on-ground gauges, radar imaging, and rainfall forecasting provided the rainfall information to drive the hydrology and hydraulic catchment modelling. As such, component 2 of the real-time procedure consisted of catchment run-off and hydraulic modeling using MIKE URBAN. The third component of the real-time procedure was to post-process the MIKE URBAN output and visualize catchment surface flooding using the MIKE DFS2 mapping capability. Each of the three components is discussed in more detail in the following sections.

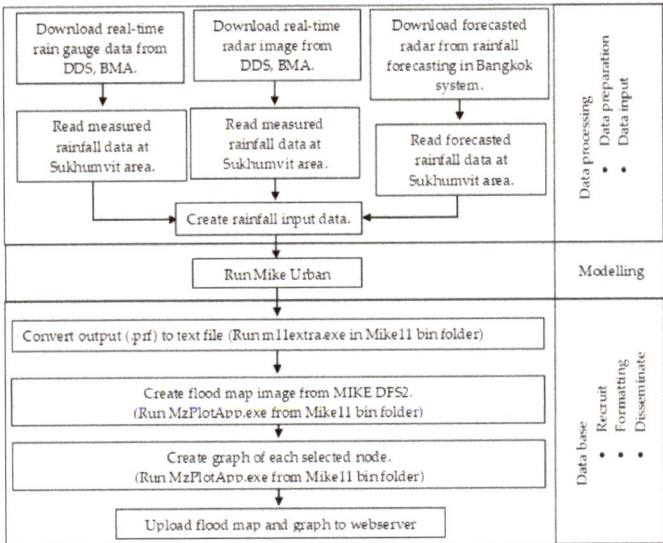

Figure 2. The real-time urban drainage system framework contains three main parts, i.e., data processing, modelling, real-time system with database.

2.3. Data Processing

2.3.1. Rainfall Input Data for Real-Time Urban Drainage System (RTUDS)

Rainfall data for the real-time urban drainage system RTUDS system were obtained from two sources, recording rain gauges and rainfall measured by radar operated by BMA, DDS.

(a) Measured Rainfall Input Data—Rain Gauges

There are five recording rain gauges (E19, E25, E26, E27, and E40) in proximity to the study area (Figure 3). The rainfall dataset period was stored from April 2014 to November 2015. The rain gauges record every 5 min, and data are available in near real-time from BMA, DDS, and the application for this RTUDS. Therefore, it was designed to use 10 min of measured rainfall input data or two time steps.

Figure 3. Location of the study area and the five rain gauge stations network which were selected for analysis in this study [19].

(b) Forecasted Rainfall Input Data—Radar

Radar images from the BMA, DDS C-Band Radar at the Nong Jok station (Figure 4), were accessed through the BMA website [26]. The range of the scanning area was 90 km, which covered 12 provinces around Bangkok. The size of the radar images was 2048 × 2034 pixels, and each pixel had an area of 0.01 sq.km. It should be noted that for this demonstration project the BMA, DDS radar was not directly accessed, as would be the case for full implementation of a real-time system, but rather the radar images were obtained directly through the BMA website.

The radar images were updated and downloaded at 5-min time steps, and following the rain gauge data procedure, two-time steps (10 min) were used to begin the forecasting process.

(c) Rainfall Forecasting

In this final rainfall processing step, the translation method [27] was applied as the main algorithm for forecasting rainfall using the radar images. From consecutive radar images, the translation program estimates translation vectors using the method of least squares, which predicts the rainfall movement pattern. After the values of parameters are determined, the current rainfall is then extrapolated into future rain. Figure 5 presents the step-by-step procedure of forecasting rainfall using the translation method [27].

Figure 4. Location of the study area and the Nong Jok radar station located at the east side of the Bangkok area which was selected for analysis in this study [26].

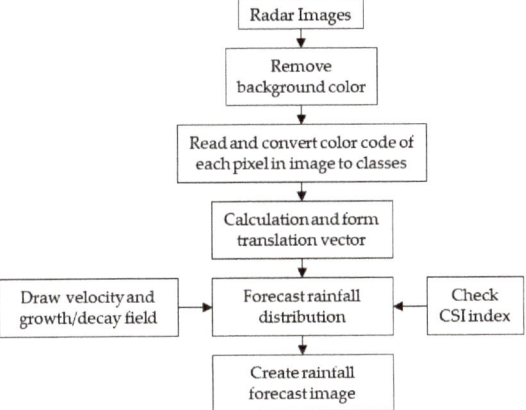

Figure 5. Flowchart of rainfall forecast using a translation method [25].

Due to the inability to directly access the radar images over the study area, the visual on-screen radar images from the BMA website, which are composed of Red (R), Green (G), and Blue (B) colors, were implemented in this study. The first step in using the radar images for the translation method was to remove all background colors other than those representing precipitation. A computer program was developed which can read the Red, Green, and Blue values in each of the pixels in a visual radar image as shown in Figure 6.

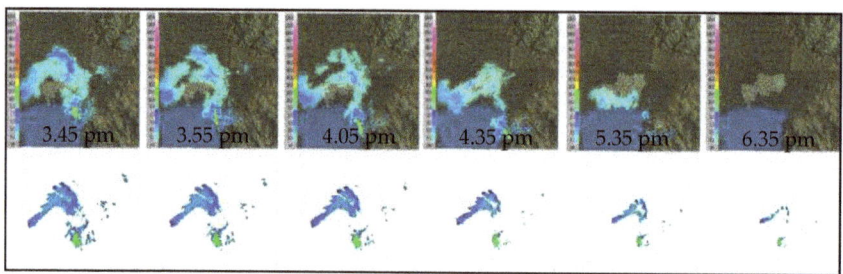

Figure 6. The result of the rainfall forecasting system for the event on 9 June 2015 at 3.35 PM [27].

There are two important factors related to rainfall data processing that will influence the efficacy of a real-time system in improving estimates of flood risk [28,29]. First, the accuracy of the rainfall forecasting system, based on the radar imagery, plays a major role in the estimated water levels produced by the catchment model. The translation forecasting system employed in this study [27] evaluated performance by the critical success index (CSI). It was used to evaluate anomalies predicted using forecasting rainfall as compared to the measured data. CSI is the ratio of the sum of hit(A), miss(B), and false(C) whereas the example outlined in the definition of CSI = (A)/(A + B + C) [30]. As in Figure 7, the optimal success of 80% occurred at a lead time of 30 min from 69 rainfall events, one of which is shown in Figure 6 as an example.

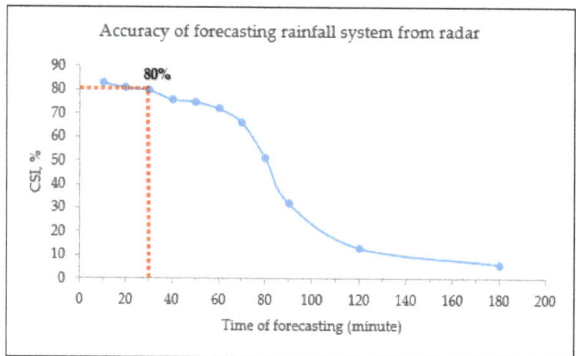

Figure 7. The accuracy of the rainfall forecasting system from radar, based on results reported in [27].

Secondly, the data processing and rainfall computation times are an essential consideration [31,32]. The rainfall forecasting system required at least two-time steps of measured rainfall data, representing a 10-min rainfall duration (i.e., two recording periods of 5 min). This study carried out a run-time analysis and determined that 40 min of the total of rainfall duration was suitable. In other words, the total of rainfall duration was a combination of 10 min of measured rainfall together with 30 min forecasted rainfall duration.

2.3.2. Urban Drainage Modelling

MIKE URBAN developed by the Danish Hydraulic Institute [24] was used to represent the overland flow flooding and pipe network hydraulics with a 1D/2D approach. Hydraulics of flow for the drainage pipe network was represented in 1D but was linked to a 2D overland flow representation for the catchment flooding. This type of linked 1D/2D approach is increasingly applied for urban flood studies as it represents more accurately the flooding of obstructions, such as buildings, although the approach is computationally more intensive than simple 1D modelling [33–36].

(a) Boundary Conditions

As noted previously, measured precipitation data and radar images were obtained from Bangkok Metropolitan Administration (BMA). The meteorological stations selected for on-ground rain gauge data in the study area were E19, E25, E26 E27, and E40 for the period of record, 2014 to 2015.

(b) Pipe Flow Model 1D

The hydraulic model of this study area developed in MIKE URBAN was based on an earlier 1D-1D approach by Chingnawan [37] in which street and sewer links were connected to each other, and the first calibration from observed data at three stations for the event of 5 October 2002 was undertaken. Validation of this model was undertaken by Chingnawan [37] with data from an event on 7 October 2002. A second calibration effort with this first-generation model was undertaken by Nguyen [25], using precipitation data and water level data from three stations for events from 16 November 2004 and 20 November 2004. The three stations that were used for the calibration of the models were Aree Station, Thonglor Station and Ekamai Station. The calibration for the 16 November 2004 event produced an efficiency index of 78%, 82%, 86%, and root mean square error of 0.06, 0.04, and 0.03 m for the three stations, respectively, while the calibration for the 24 November 2004 resulted in an efficiency index of 78%, 82%, 86% and root mean square error of 0.06, 0.04 and 0.03 m for the three stations, respectively. In 2013, the model was revised by Shrestha [38] using a coupled 1D/2D approach. This second-generation 1D/2D model was calibrated using the measured water level and rainfall data for the storm event of 15 October 2003. The Sukhumvit 26 and Ekamai water level stations were used in the calibration, with the root mean square error (RMSE) and efficiency index (EI) for the Sukhumvit 26 station being 0.04 m and 82%, respectively, and the Ekamai station having RMSE and EI values of 0.24 and 60%, respectively. The results for these first- and second-generation models indicate that MIKE URBAN can be applied with relatively good accuracy for the purposes of this current study.

(c) Overland Flow Model 2D

The second-generation model developed by Shrestha [38] served as the basis for this study, but all model components such as pipe network, pumping station, and outlet configurations were updated. The overland flow model in the current study was represented using a DEM. The elevation gradient in the grid cells defines the topography of the study area. The computation grid size had a 10 m resolution and a computation time step of 2 sec to ensure model stability. All the nodes except outlets were coupled to 2D cells at the ground elevation, as in Figure 8. The characteristics of the drainage system and typologies of roads and buildings have a substantial impact on the water level in flooded areas [39]. Those points were applied to each component of 2D hydraulic model, such as pipe network in node and link, road and building as 2D cells for overland flow simulation. In this study, the validation of the 2D hydrodynamic model was performed by comparing modeled flood areas produced by MIKE FLOOD with a map of flood-prone areas provided by BMA. In addition, an assessment of the overland model was carried out using a condition of 45% imperviousness per the study by Shrestha [38] with measured water level and rainfall data for the event of 11 October 2014. For this event, the flood risk areas confirmed by the BMA matched almost identically to the map of their flood-prone area. The overland flow model for the 11 October 2014 event from 11:50 AM to 2:00 PM is summarized in Figure 9. From the flood map and model results (Figure 9), the maximum floodwater depth for the rainfall event varied between 0.025 m to 0.125 m. Compared to the map of flood-prone areas developed by the DDS, BMA in Figure 10, almost the same streets exhibit a flooding pattern. In this way, the validation of the 1D-2D model can be considered as excellent for the Sukhumvit area at a planning scale.

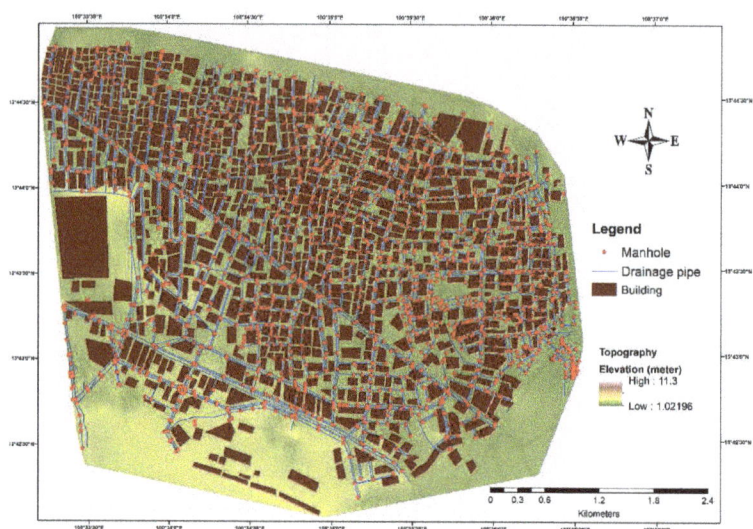

Figure 8. Overland flow of the coupled 1D/2D model at Sukhumvit area. The map show network of the drainage pipes and digital elevation model (DEM) which provide topographics of the study area.

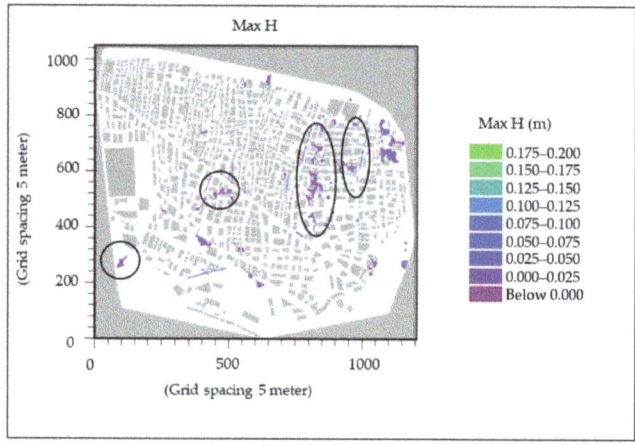

Figure 9. The maximum flood depth (Max H) from 2D overland flow model on 11 October 2014.

The overland flow model was validated using the rainfall data on 11 October 2014 from 11:50 AM to 2:00 PM. The result of the model was compared with the observed flood data on the day, as shown in Figure 9.

From the flood map and model results (Figure 9), the maximum floodwater depth for the rainfall event on 11 October 2014 varies between 0.025 m to 0.125 m. Compared the flood map to the flood risk areas map developed by the DDS, BMA in Figure 10, almost the same streets exhibit a flooding pattern. In this way, the validation of the 1D-2D model can be considered as excellent for the Sukhumvit area.

Figure 10. Flood risk map showing the risk area by yellow line in Sukhumvit polder area. This map is from the statistical profile of Bangkok Metropolitan Administration report [19].

2.3.3. The Real-Time Urban Drainage System Design

(a) Analysis of Forecasted Lead Time

The model run time was tested using different cases that varied duration of measured and forecasted rainfall. The shortest period of rainfall was from both rain gauges and radar at 5 min time steps. The computation time recorded from the start running time until the model finishes running. To set up running time for the system, there were several different cases of rainfall duration starting from the minimum time step of rainfall data that was 5 min to at least 60 min. Moreover, from rainfall duration, it divided into measured and forecast rainfall duration, which was the total rainfall duration of the system. To set the overall decision-making time, the other task processing times also needed to be considered.

These considerations can be an important limitation of setting up the real-time system. If the model run time and overall decision-making time is too long, the advantage of real-time modelling is lost. Moreover, the radar forecast needs at least three recorded images in order to forecast the rainfall. Thus, the measured rainfall duration will be 10 min, and the rest of the duration will be the forecast and the decision part, which takes 30 min. The model running time plus other tasks will take 20 min, and there are around 10 min of lead time from each forecasting time of the RTUDS.

(b) Flood Forecast Dissemination

The presentation of time-series water level and inundation map and animation are produced to help flood agencies and the community to understand the flood risk for the area and to manage and decrease flood damage as the result real-time actions that may be implemented. The RTUDS results are presented in two ways. The first method reports the MIKE URBAN simulation results via Mzplotcomapp.exe which visualizes the time series of water levels at various considered nodes. The monitor stations of public services such as school and hospital, and the criteria of selected locations that will show the time series of water level. The simulated results will be concentrated on the area in which flooding always occurs. All the areas having time series of water level results will be shown automatically, and the flood map presentation has the benefit of being understood easily and of timely implementation for all urban flooding.

3. Results

Model Accuracy and Limitation

A real-time urban drainage system was successfully developed in this study. The model running time was tested using different cases that varied the duration of rainfall. The shortest period of precipitation was 40 min. The time recorded from the start running time until the model finish running was 20 min. It covered modelling time and the other tasks such as computing, sending, and receiving data. The 40 min of total rainfall duration that input to the model will contain 10 min of measured rainfall duration and 30 min of forecasted duration. The lead time of the forecast is approximately 10 min ahead due to current computer specification as shown in Figure 11.

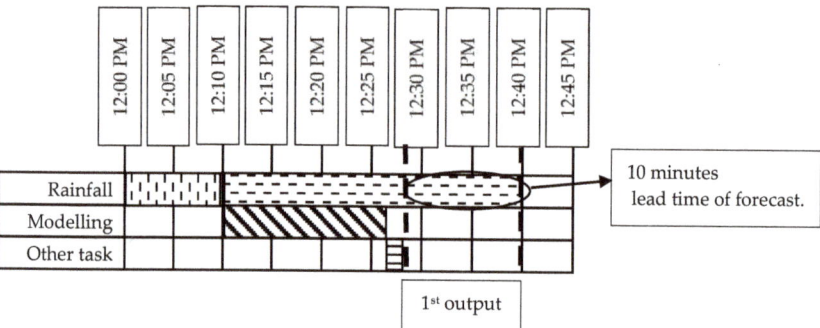

Figure 11. The real-time urban drainage system timeline, explaining each part of real-time system such as rainfall, modelling and post processing before display results.

Figure 12 shows the results of the simulated water level at two selected nodes and the maximum flood level map in the study with three different sources of rainfall data.

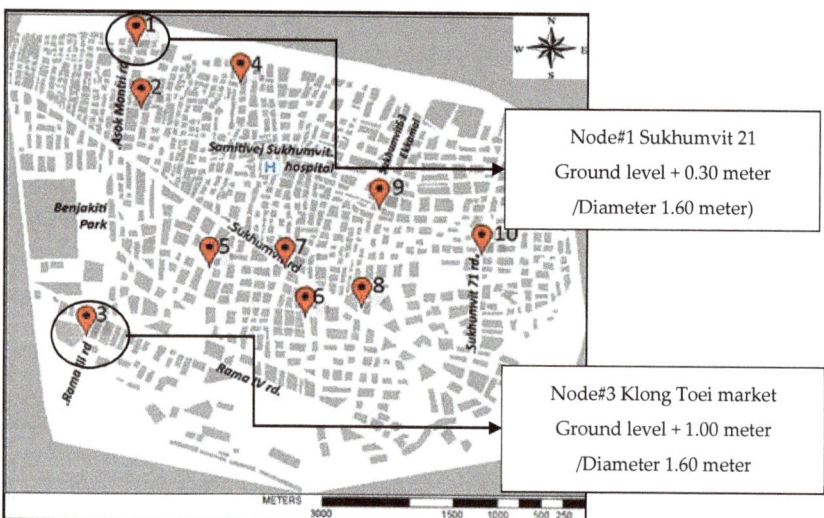

Figure 12. The map shows selected locations of Node 1 and 3 for comparing the water level results from the various sources of rainfall data.

This study consists of the simulation results based on 10 rainfall events on 8 June 2015 from 0.00 AM to 6.00 AM. It is divided into three different scenarios of rainfall input sources such as:

(1) Rainfall from RTUDS

In this case, the real-time system used 40 min of rainfall from a combination of 10 min of recorded rainfall and 30 min of forecasted rainfall. It showed the continuous results of 10 events, as shown in Figures 13 and 14.

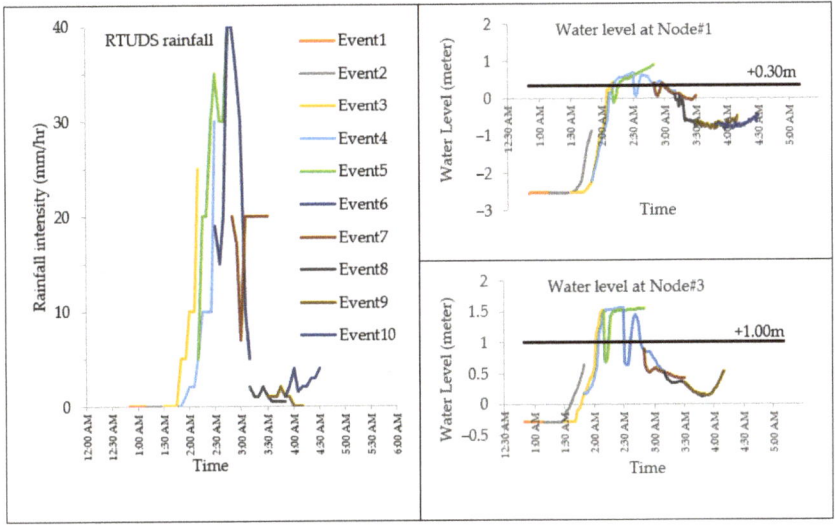

Figure 13. Rainfall and water level from real-time urban drainage system rainfall.

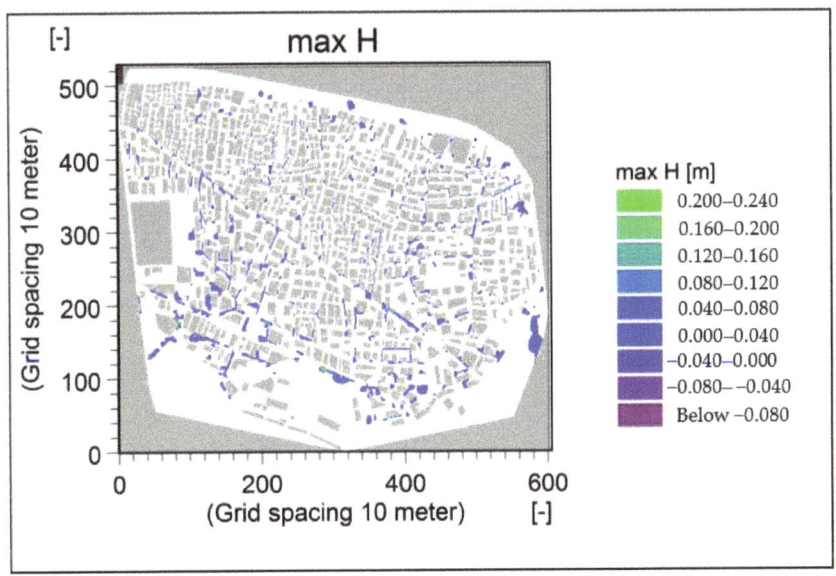

Figure 14. Maximum flood map from real-time urban drainage system rainfall.

(2) Rainfall from Measured Radar

In this case, the result of water level and flood inundation was produced to compare with RTUDS. The rainfall input was used from measured radar data with the same 40 min in each rainfall event for simulation (Figures 15 and 16).

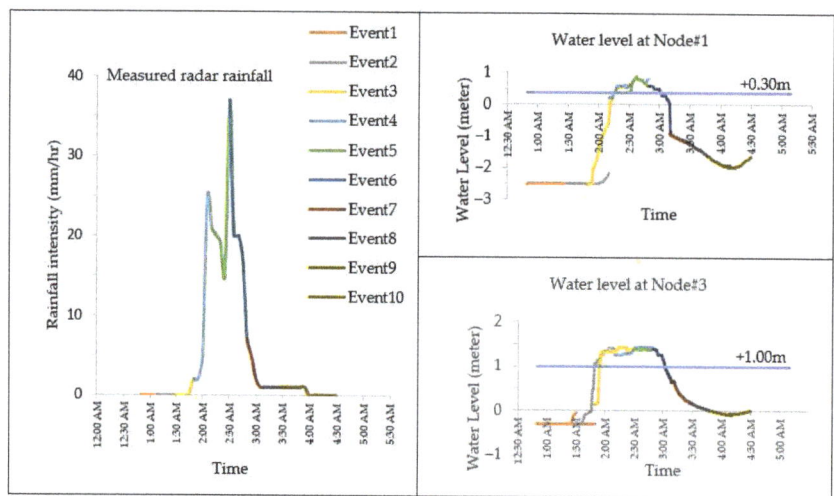

Figure 15. Rainfall and water level from measured radar rainfall.

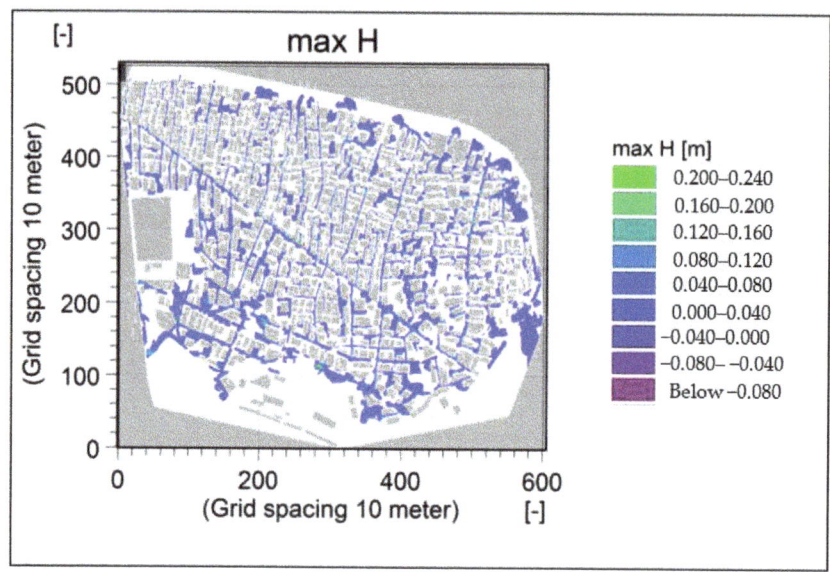

Figure 16. Maximum flood map result from measured radar rainfall.

(3) Rainfall from Rain Gauge Station

This case used measured rainfall from the rain gauge station at station E40 as a represent station of the area. The rainfall duration was 40 min in each rainfall event for simulation (Figures 17 and 18).

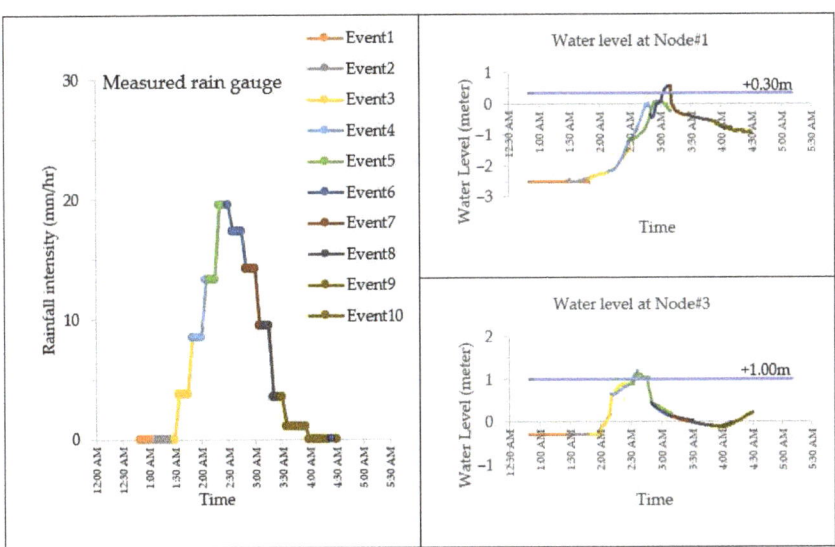

Figure 17. Rainfall and water level from measured rain gauge.

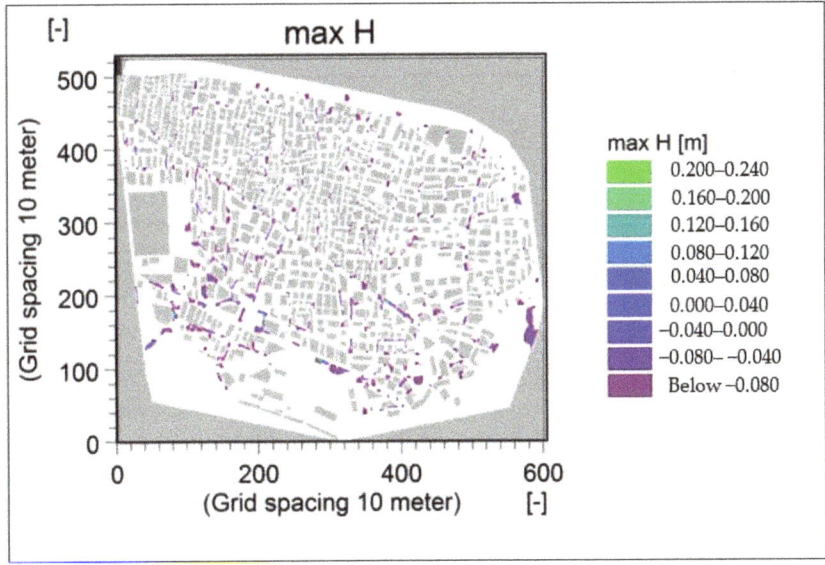

Figure 18. Maximum flood map result from the measured rain gauge.

- Real-Time Urban Drainage System—RTUDS Rainfall
- Measured radar rainfall.
- Measured rain gauge.

The RTUDS provided the maximum depth of flood, and the notification flood occurs from three different rainfall source scenarios. Table 1 summarizes the results.

Table 1. Comparison of flood notification from the result of real-time urban drainage system (RTUDS) and measured rainfall on 8 June 2015 from 0.00 AM to 6.00 AM.

Event No.	Model Start Running at	Rainfall Duration	Flood Notification		
			RTUDS	Measured Radar	Measured Rain Gauge
1	1:00 AM	0:50–1:30 AM	-	-	-
2	1:20 AM	1:10–1:50 AM	-	-	-
3	1:40 AM	1:30–2:10 AM	Flood	-	-
4	2:00 AM	1:50–2:30 AM	Flood	Flood	-
5	2:20 AM	2:10–2:50 AM	Flood	Flood	-
6	2:40 AM	2:30–3:10 AM	Flood	Flood	Flood
7	3:00 AM	2:50–3:30 AM	-	Flood	Flood
8	3:20 AM	3:10–3:50 AM	-	-	Flood
9	3:40 AM	3:30–4:10 AM	-	-	-
10	4:00 AM	3:50–4:30 AM	-	-	-

On the same day, BMA, DDS reported flooding occurred from 3.00 AM, and covered most of the main roads in study areas such as Rama III, IV, Sukhumvit, and Asokemontri. The results of the analysis here illustrate that the real-time urban drainage system can provide advanced notice of 10 min of the potential for flooding. The advanced notice provided by the RTUDS can facilitate preparation time for BMA DDS to manage the system through the operation of flood gates and deployment of pumps to critical areas. In addition, people in the area can be provided with ongoing updates of flood status and risk. Moreover, RTUDS rainfall is based on both measured and forecasted radar rainfall. Therefore, the accuracy depends on the forecasting of radar rainfall data. The lead time of the forecast of around 10 min to know the future situation is still quite short. It needs to be improved to make the lead time of the forecast longer.

4. Discussion

Previous studies undertaken in Sukhumvit [25,37] have shown the significance of real-time system accuracy, and the limitations based on the forecasting methods and the computer limitation. This study has achieved the 10 min lead time by using the present real-time modeling with 40 min rainfall duration and 15–20 min processing time. This result has greatly improved on previous research [25]. It showed that there is a significant increase in term of processing frequency duration from 60 min of rainfall duration with 60 min of processing time to 40 min of rainfall duration with 15–20 min of processing time due to the computer limitation and the updated time series of the rainfall data source. Although the lead time will be short, this study can provide the forecasting results over a shorter duration. Therefore, people or related officers can use these results for deciding to avoid flooding. In addition, computer performance is one of the most important parts of the real-time system. Therefore, computational time will be the main parameter to design the schedule of the system and determine the lead time of the forecast. The technology of software for modeling and hardware in terms of computer devices has been developed from previous studies until now. Thus, computation time in this study is less than in the past, and it allows the real-time forecasting model to provide a longer lead time of forecasting and more frequent results.

5. Conclusions

The study highlighted the challenges of real-time urban flood warning development. The methodology of automatic processing can be applied to other areas of urban and peri-urban Bangkok and has potential to be developed for other cities in Southeast Asia. The integrated system-linked rainfall forecasting from radar, drainage pipe flow modelling and the overland flood modelling in an effort to streamline the flood risk prediction process.

In this way, the research can help to reduce the complicated process from computing and provide shorter and more frequent flood forecast updates.

The selected 40 min of rainfall data (10 min record and 30 min forecast) was optimally applied in the real-time system. The system updates every 20 min and provides the lead time forecast of 10 min. This is a computation and analysis improvement compared to the one hour of time needed for flood prediction as reported by Boonya-aroonnet [40] for a single event. However, the improvement of this study came from the integration of software contained with rainfall forecasting and flood simulation. In addition, this study benefited from improved hardware and software technology as compared to the early 2000s. The shorter simulation time and more frequent reporting of results can be further developed for a flood warning system.

Moreover, the real-time urban drainage system still needs a stable source of rainfall data. Otherwise, it will be terminated. The automatic switch mode to choose the source of the rainfall data can help the system run continuously. The boundary around the study area must be linked to the model in real-time—for example, the water level in the canal around the city. The future technology of computer servers and databases can reduce the complex steps of the computer system and shorten the model running time. More capacity of the memory in the computer will create more space to store the backup result file. Otherwise, it needs to be taken out and stored in external memory or on the cloud. Additional field measurement is recommended. For example, the discharge of the sewer should be measured. The comparison of the calculating and field measurement discharge is essential to obtain higher model accuracy.

Author Contributions: Conceptualization, D.C. and S.W.; methodology, D.C.; software, D.C.; validation, D.C.; formal analysis, D.C.; investigation, D.C.; data curation, D.C.; writing—original draft preparation, D.C.; writing—review and editing, D.C. and H.M.; visualization, D.C. and H.M.; supervision, S.W. and H.M. All authors have read and agreed to the published version of the manuscript.

Funding: This research received no external funding.

Institutional Review Board Statement: Not applicable.

Informed Consent Statement: Not applicable.

Data Availability Statement: Not applicable.

Acknowledgments: The authors wish sincere thanks to the Drainage and Sewerage Department, the Bangkok Metropolitan Administration for providing data and necessary information for the research. The authors also would like to thank all laboratory members for their advice and helps in the computer technical. Finally, we would like to thank three anonymous reviewers and editors for thorough and constructive comments that greatly improved the clarity and impact of this work.

Conflicts of Interest: The authors declare no conflict of interest.

References

1. René, J.-R.; Djordjević, S.; Butler, D.; Madsen, H.; Mark, O. Assessing the potential for real-time urban flood forecasting based on a worldwide survey on data availability. *Urban Water J.* **2013**, *11*, 573–583. [CrossRef]
2. Irvine, K.N. Climate Change and Urban Hydrology: Research Needs in the Developed and Developing Worlds. *J. Water Manag. Model.* **2013**. [CrossRef]
3. Chang, C.-H.; Irvine, K. Climate Change Resilience and Public Education in Response to Hydrologic Extremes in Singapore. *Br. J. Environ. Clim. Chang.* **2014**, *4*, 328–354. [CrossRef]
4. Ruangpan, L.; Vojinovic, Z.; Di Sabatino, S.; Leo, L.S.; Capobianco, V.; Oen, A.M.P.; McClain, M.E.; Lopez-Gunn, E. Nature-based solutions for hydro-meteorological risk reduction: A state-of-the-art review of the research area. *Nat. Hazards Earth Syst. Sci.* **2020**, *20*, 243–270. [CrossRef]
5. Zevenbergen, C.; Fu, D.; Pathirana, A. Transitioning to Sponge Cities: Challenges and Opportunities to Address Urban Water Problems in China. *Water* **2018**, *10*, 1230. [CrossRef]
6. Fletcher, T.D.; Shuster, W.; Hunt, W.F.; Ashley, R.; Butler, D.; Arthur, S.; Trowsdale, S.; Barraud, S.; Semadeni-Davies, A.; Bertrand-Krajewski, J.-L.; et al. SUDS, LID, BMPs, WSUD and more—The evolution and application of terminology surrounding urban drainage. *Urban Water J.* **2014**, *12*, 525–542. [CrossRef]

7. Irvine, K.N.; Loc, H.H.; Sovann, C.; Suwanarit, A.; Likitswat, F.; Jindal, R.; Koottatep, T.; Gaut, L.H.C.; Qi, L.W.; Wandeler, K.D. Bridging the Form and Function Gap in Urban Green Space Design through Environmental Systems Modeling. *J. Water Manag. Model.* **2021**. [CrossRef]
8. Lashford, C.; Rubinato, M.; Cai, Y.; Hou, J.; Abolfathi, S.; Coupe, S.; Charlesworth, S.; Tait, S. SuDS & Sponge Cities: A Comparative Analysis of the Implementation of Pluvial Flood Management in the UK and China. *Sustainability* **2019**, *11*, 213. [CrossRef]
9. Lim, H.; Lu, X. Sustainable urban stormwater management in the tropics: An evaluation of Singapore's ABC Waters Program. *J. Hydrol.* **2016**, *538*, 842–862. [CrossRef]
10. Kang, N.; Kim, S.; Kim, Y.; Noh, H.; Hong, S.J.; Kim, H.S. Urban Drainage System Improvement for Climate Change Adaptation. *Water* **2016**, *8*, 268. [CrossRef]
11. Oladunjoye, O.A.; Proverbs, D.G.; Collins, B.; Xiao, H. A cost-benefit analysis model for the retrofit of sustainable urban drainage systems towards improved flood risk mitigation. *Int. J. Build. Pathol. Adapt.* **2019**, *38*, 423–439. [CrossRef]
12. Ngo, T.T.; Jung, D.; Kim, J.H. Robust Urban Drainage System: Development of a Novel Multiscenario-Based Design Approach. *J. Water Resour. Plan. Manag.* **2019**, *145*, 04019027. [CrossRef]
13. Nazari, B.; Seo, D.-J.; Muttiah, R.; Worth, T.C.O.F. Assessing the Impact of Variations in Hydrologic, Hydraulic and Hydrometeo-rological Controls on Inundation in Urban Areas. *J. Water Manag. Model.* **2016**. [CrossRef]
14. Kim, H.I.; Keum, H.J.; Han, K.Y. Real-Time Urban Inundation Prediction Combining Hydraulic and Probabilistic Methods. *Water* **2019**, *11*, 293. [CrossRef]
15. Chang, F.-J.; Chen, P.-A.; Lu, Y.-R.; Huang, E.; Chang, K.-Y. Real-time multi-step-ahead water level forecasting by recurrent neural networks for urban flood control. *J. Hydrol.* **2014**, *517*, 836–846. [CrossRef]
16. Mason, D.; Giustarini, L.; Garcia-Pintado, J.; Cloke, H. Detection of flooded urban areas in high resolution Synthetic Aperture Radar images using double scattering. *Int. J. Appl. Earth Obs. Geoinf.* **2014**, *28*, 150–159. [CrossRef]
17. Rafieeinasab, A.; Norouzi, A.; Kim, S.; Habibi, H.; Nazari, B.; Seo, D.-J.; Lee, H.; Cosgrove, B.; Cui, Z. Toward high-resolution flash flood prediction in large urban areas—Analysis of sensitivity to spatiotemporal resolution of rainfall input and hydrologic modeling. *J. Hydrol.* **2015**, *531*, 370–388. [CrossRef]
18. Loc, H.H.; Park, E.; Chitwatkulsiri, D.; Lim, J.; Yun, S.-H.; Maneechot, L.; Phuong, D.M. Local rainfall or river overflow? Re-evaluating the cause of the Great 2011 Thailand flood. *J. Hydrol.* **2020**, *589*, 125368. [CrossRef]
19. Drainage and Sewerage Department. Available online: https://dds.bangkok.go.th/content/doc3/index.php (accessed on 28 March 2021).
20. Laeni, N.; Brink, M.V.D.; Arts, J. Is Bangkok becoming more resilient to flooding? A framing analysis of Bangkok's flood resilience policy combining insights from both insiders and outsiders. *Cities* **2019**, *90*, 157–167. [CrossRef]
21. Hilly, G.; Vojinovic, Z.; Weesakul, S.; Sanchez, A.; Hoang, D.N.; Djordjević, S.; Chen, A.S.; Evans, B. Methodological Framework for Analysing Cascading Effects from Flood Events: The Case of Sukhumvit Area, Bangkok, Thailand. *Water* **2018**, *10*, 81. [CrossRef]
22. Nabangchang, O.; Allaire, M.; Leangcharoen, P.; Jarungrattanapong, R.; Whittington, D. Economic costs incurred by households in the 2011 Greater Bangkok flood. *Water Resour. Res.* **2014**, *51*, 58–77. [CrossRef]
23. Weesakul, S.; Mark, O.; Naksua, W.; Chingnawan, S.; Liong, S.-Y.; Phoon, K.-K.; Babovic, V. Real time urban flood modeling for bangkok metropolitan administration, case study: Sukumvit area. In *Hydroinformatics*; World Scientific Publishing: Singapore, 2004; pp. 1907–1914.
24. Mike. Flood. Available online: https://www.mikepoweredbydhi.com/products/mike-flood (accessed on 28 March 2021).
25. Hung, M.E.N.Q.; Weesakul, S.; Weesakul, U.; Chaliraktrakul, C.; Mark, O.; Larsen, L.C. A Real-Time Hydrological Information System for Cities. *Water Encycl.* **2005**, 121–127. [CrossRef]
26. Weather Bangkok. Available online: http://weather.bangkok.go.th/radar/RadarNongchok.aspx (accessed on 28 March 2021).
27. Weesakul, S. Development of Rainfall Forecasting Platform in Bangkok. Available online: https://www.nstda.or.th/th/2436-rainalert (accessed on 16 December 2020).
28. Toth, E.; Brath, A.; Montanari, A. Comparison of short-term rainfall prediction models for real-time flood forecasting. *J. Hydrol.* **2000**, *239*, 132–147. [CrossRef]
29. Henonin, J.; Russo, B.; Mark, O.; Gourbesville, P. Real-time urban flood forecasting and modelling—A state of the art. *J. Hydroinform.* **2013**, *15*, 717–736. [CrossRef]
30. Shah, R.; Sahai, A.K.; Mishra, V. Short to sub-seasonal hydrologic forecast to manage water and agricultural resources in India. *Hydrol. Earth Syst. Sci.* **2017**, *21*, 707–720. [CrossRef]
31. Jeong, J.; Kannan, N.; Arnold, J.; Glick, R.H.; Gosselink, L.; Srinivasan, R. Development and Integration of Sub-hourly Rainfall–Runoff Modeling Capability Within a Watershed Model. *Water Resour. Manag.* **2010**, *24*, 4505–4527. [CrossRef]
32. Notaro, V.; Fontanazza, C.; Freni, G.; Puleo, V. Impact of rainfall data resolution in time and space on the urban flooding evaluation. *Water Sci. Technol.* **2013**, *68*, 1984–1993. [CrossRef]
33. Leandro, J.; Chen, A.S.; Djordjević, S.; Savic, D. Comparison of 1D/1D and 1D/2D Coupled (Sewer/Surface) Hydraulic Models for Urban Flood Simulation. *J. Hydraul. Eng.* **2009**, *135*, 495–504. [CrossRef]
34. Vojinovic, Z.; Tutulic, D. On the use of 1D and coupled 1D-2D modelling approaches for assessment of flood damage in urban areas. *Urban Water J.* **2009**, *6*, 183–199. [CrossRef]

35. Marvin, J.T.; Wilson, A.T. One Dimensional, Two Dimensional and Three Dimensional Hydrodynamic Modeling of a Dyked Coastal River in the Bay of Fundy. *J. Water Manag. Model.* **2016**. [CrossRef]
36. Abdelrahman, Y.T.; El Moustafa, A.M.; Elfawy, M. Simulating Flood Urban Drainage Networks through 1D/2D Model Analysis. *J. Water Manag. Model.* **2018**. [CrossRef]
37. Chingnawan, S. *Real-Time Modelling of Urban Flooding in the Sukhumvit Area, Bangkok, Thailand*; Asian Institute of Technology: Phatumthani, Thailand, 2003.
38. Shrestha, A.; Babel, M.S.; Weesakul, S. Integrated Modelling of Climate Change and Urban Drainage. *Manag. Water Resour. Under Clim. Uncertain.* **2014**, 89–103. [CrossRef]
39. Postacchini, M.; Zitti, G.; Giordano, E.; Clementi, F.; Darvini, G.; Lenci, S. Flood impact on masonry buildings: The effect of flow characteristics and incidence angle. *J. Fluids Struct.* **2019**, *88*, 48–70. [CrossRef]
40. Boonya-Aroonnet, S.; Weesakul, S.; Mark, O. Modeling of Urban Flooding in Bangkok. *Glob. Solut. Urban Drain.* **2002**, 1–14. [CrossRef]

MDPI
St. Alban-Anlage 66
4052 Basel
Switzerland
Tel. +41 61 683 77 34
Fax +41 61 302 89 18
www.mdpi.com

Water Editorial Office
E-mail: water@mdpi.com
www.mdpi.com/journal/water

www.ingramcontent.com/pod-product-compliance
Lightning Source LLC
LaVergne TN
LVHW070725100526
838202LV00013B/1175